Coastal Morphology Assessment and Coastal Protection

Editor

Yoshimichi Yamamoto

MDPI • Basel • Beijing • Wuhan • Barcelona • Belgrade • Manchester • Tokyo • Cluj • Tianjin

Editor
Yoshimichi Yamamoto
Department of Civil Engineering
Tokai University
Hiratsuka
Japan

Editorial Office
MDPI
St. Alban-Anlage 66
4052 Basel, Switzerland

This is a reprint of articles from the Special Issue published online in the open access journal *Journal of Marine Science and Engineering* (ISSN 2077-1312) (available at: www.mdpi.com/journal/jmse/special_issues/coastal_morphology).

For citation purposes, cite each article independently as indicated on the article page online and as indicated below:

LastName, A.A.; LastName, B.B.; LastName, C.C. Article Title. *Journal Name* **Year**, *Volume Number*, Page Range.

ISBN 978-3-0365-1681-3 (Hbk)
ISBN 978-3-0365-1682-0 (PDF)

Contents

Preface to "Coastal Morphology Assessment and Coastal Protection"

It has been a long time since excessive industrial activities by humans have had many negative effects on the natural environment. Dredging for aggregate collection and waterway maintenance has reduced the supply of sediment to coasts, and dams and breakwaters have side effects that impede the supply of sediment to coasts, resulting in serious coastal erosion. In addition, the generation of large amounts of greenhouse gases is causing sea level rise and intensification of typhoons. Furthermore, huge earthquakes and tsunamis have often occurred in the Pacific coastal areas, but the scale of damage caused by these has increased due to the excessive concentration of population and assets in the coastal areas.

Therefore, In order to reduce coastal disasters, it is very important to elucidate phenomena related to coastal disasters by improving various observation techniques and analysis methods, and to continue improving prediction methods for diverse disaster phenomena.

I hope this book helps you with this purpose.

Yoshimichi Yamamoto
Editor

Journal of
Marine Science and Engineering

MDPI

Editorial

Coastal Morphology Assessment and Coastal Protection

Yoshimichi Yamamoto

Department of Civil Engineering, Tokai University, Kanagawa 259-1292, Japan; yo-yamamo@tokai.ac.jp

check for
updates

Citation: Yamamoto, Y. Coastal Morphology Assessment and Coastal Protection. *J. Mar. Sci. Eng.* **2021**, *9*, 713. https://doi.org/10.3390/jmse9070713

Received: 2 June 2021
Accepted: 7 June 2021
Published: 28 June 2021

Publisher's Note: MDPI stays neutral with regard to jurisdictional claims in published maps and institutional affiliations.

1. Introduction

Sediment-collecting in rivers and seas to secure a large amount of aggregate reduces the supply of earth and sand to coasts. Dams and breakwaters constructed in various places impede the transportation of earth and sand as a side effect. Furthermore, the maintenance dredging of dam lakes and waterways will also disrupt the supply of sediment to coasts if the dredged sediment is not released back to the water system. Due to these development activities, coastal erosion has become a serious problem on many beaches around the world. Moreover, due to excessive industrial activities by human beings, the exacerbation of natural disasters caused by global warming is becoming a real problem. In addition, because great earthquakes with a magnitude of 9 or more have occurred about three times per 100 years at boundaries of the Pacific Crust Plate and the Nazca Crust Plate since 1700, the possibility of losing many lives and assets in the Pacific coastal areas due to a huge tsunami caused by a great earthquake should also not be underestimated. Therefore, research into the prevention and mitigation of coastal erosion and coastal disasters is becoming increasingly important.

In this Special Issue, five studies on coastal erosion and coastal disaster prevention resulting from large waves and tsunamis were collected to contribute to the technological progress in this field.

2. Coastal Beach Change by Large Waves

Since around 1960, when the coastal erosion problem became more serious, many researchers have participated in research into beach topography changes, and many practical numerical models for forecasting beach deformation have emerged. (XBeach model [1], DELFT3D Model [2], 3D Coupled Fluid-Structure-Sediment Interaction Model [3], etc.). However, when performing a long-term prediction using a 3D or horizontal 2D numerical model, if the external force for the entire period is input, the calculation time becomes very long. Therefore, it is necessary to select a limited number of external forces that are dominant against topographical changes to save calculation time. From this perspective, it is important to elucidate the topographical change phenomenon of a target coast from the analysis using long-term information on topographical changes and external forces. Nuyts et al. [4] clarified the trends of topographical changes in the Rossbeigh coastal barrier in Dingle Bay, Ireland, using detailed analyses of 19-year data from the global navigation satellite system, bathymetry surveying, and an external force observation system. Moreover, Kelpšaitė-Rimkienė et al. [5] revealed the characteristics of the cross-shore profile change in the Palanga coast of the East Baltic Sea through the skillful analysis of topographic survey data and wind data using K-means Clustering Technique, and calculation of the cross-shore sediment transport rate.

Furthermore, since gravel barrier beaches are important as natural breakwaters, Ions et al. [6] predicted the amount of erosion and overwash per unit length on the beach, caused by important combinations of external forces using the X Beach model on the gravel barrier beach of Hurst Castle Spit in UK. Then, they organized these simulated values by the barrier inertia parameter (=the pre-storm barrier cross sectional area × the initial barrier freeboard/the cube of the incident wave height) and created calculation charts to obtain the amount of erosion and overwash per unit length.

1

3. Coastal Disasters by Tsunamis

Calculation methods for tsunami inundation and topographical changes due to tsunamis has also progressed rapidly since around 1990, when data began to increase due to the spread of observation technology, and 3D prediction numerical models have emerged (DELFT3D model [2], 3D Coupled Fluid-Structure-Sediment Interaction Model [7], etc.). However, due to the large geographic areas which need to be modelled, the computational costs of 3D tsunami simulations can be extremely high. Therefore, simulations using an ordinary computer becomes infeasible. Ahmadi et al. [8] improved the prediction method technique using a numerical model of horizontal 2D inundation and topographical change that can be executed by an ordinary computer. The main sediment movement of topographical changes is "bed load", and the coefficients of formulae to calculate the bed load must be determined in advance by verification simulations based on actual measurement data of the target coast. However, without the actual measurement data, it became difficult to numerically reproduce topographical changes. Therefore, they conducted vertical 2D hydraulic experiments with models that were as large as possible to create calculation charts that can easily determine the coefficient of a formula for calculating the bed load. In addition, they proposed diagrams that can easily determine the threshold value of the width of columns, where individual buildings collapse due to the tsunami, using inundation depth data obtained from the calculation results of the horizontal 2D numerical model.

In the case of a huge tsunami, because the inundation area becomes vast and the amount of generated drifting objects become enormous, it is also important to provide methods that can correctly evaluate the impact force when a drifting object hits a structure. Therefore, Yamamoto et al. [9] proposed appropriate impact force calculation formulae for each type of drifting object, based on the correlation examination using the impact force data from collision experiments with large models.

4. Conclusions

This Special Issue has introduced two analysis methods using the data of coastal topography changes and external forces, which are useful for grasping the characteristics of topography changes due to large waves and examining coastal protection measures, and an effective use method of the Xbeach model to obtain calculation charts that contribute to coastal protection. Moreover, this Special Issue has introduced a method that can predict the wide-area topographical changes and building destruction caused by a huge tsunami relatively easily, and appropriate formulae for estimating the impact force for each driftage type, in order to prevent the structural destruction caused by drifting objects generated in large numbers by the huge tsunami. The authors sincerely hope that this Special Issue will be useful for coastal disaster prevention.

Funding: This editorial received no external funding.

Institutional Review Board Statement: Not applicable.

Informed Consent Statement: Not applicable.

Acknowledgments: All contributing authors and reviewers are thanked for their efforts.

Conflicts of Interest: The author declared no conflict of interest.

References

1. UNESCO-IHE, Deltares, Delft University of Technology, and University of Miami. Available online: https://oss.deltares.nl/web/xbeach/home (accessed on 25 June 2021).
2. Deltares, and Delft University of Technology. Available online: https://oss.deltares.nl/web/delft3d (accessed on 25 June 2021).
3. Tomoaki, N.; Solomon, C.Y.; Norimi, M. Development of Three-Dimensional Coupled Fluid-Structure-Sediment Interaction Model and its Application to Local Scour around Submerged Object. *J. Jpn. Soc. Civ. Eng. B2* **2010**, *66*, 406–410. (In Japanese)
4. Siegmund, N.; Michael, O.; Jimmy, M. Monitoring the Morphodynamic Cannibalization of the Rossbeigh Coastal Barrier and Dune System over a 19-Year Period (2001–2019). *J. Mar. Sci. Eng.* **2020**, *8*, 421. [CrossRef]

5. Kelpšaitė-Rimkienė, L.; Parnell, K.E.; Žaromskis, R.; Kondrat, V. Cross-Shore Profile Evolution after an Extreme Erosion Event—Palanga, Lithuania. *J. Mar. Sci. Eng.* **2021**, *9*, 38. [CrossRef]
6. Ions, K.; Karunarathna, H.; Reeve, D.E.; Pender, D. Gravel Barrier Beach Morphodynamic Response to Extreme Conditions. *J. Mar. Sci. Eng.* **2021**, *9*, 135. [CrossRef]
7. Nakamura, T.; Mizutani, N. Numerical Analysis of Tsunami Overflow over a Coastal Dike and Local Scouring at the Toe of its Landward Slope. *J. Jpn. Soc. Civ. Eng. B3* **2014**, *70*, I_516–I_521. (In Japanese)
8. Ahmadi, S.M.; Yamamoto, Y.; Ca, V.T. Rational Evaluation Methods of Topographical Change and Building Destruction in the Inundation Area by a Huge Tsunami. *J. Mar. Sci. Eng.* **2020**, *8*, 762. [CrossRef]
9. Yamamoto, Y.; Kozono, Y.; Mas, E.; Murase, F.; Nishioka, Y.; Okinaga, T.; Takeda, M. Applicability of Calculation Formulae of Impact Force by Tsunami Driftage. *J. Mar. Sci. Eng.* **2021**, *9*, 493. [CrossRef]

Journal of
Marine Science and Engineering

Article

Monitoring the Morphodynamic Cannibalization of the Rossbeigh Coastal Barrier and Dune System over a 19-Year Period (2001–2019)

Siegmund Nuyts, Michael O'Shea * and Jimmy Murphy

Environmental Research Institute, Beaufort Building, MaREI Centre, Cork P43 C573, Ireland;
Siegmund.nuyts@ucc.ie (S.N.); jimmy.murphy@ucc.ie (J.M.)
* Correspondence: michaeloshea@ucc.ie

Received: 25 May 2020; Accepted: 7 June 2020; Published: 9 June 2020

Abstract: This research presents a study on the morphodynamic evolution of the Rossbeigh coastal barrier and its dune system, located in Dingle Bay, County Kerry, Ireland. The study examines the evolution of the system over a 19-year period (2001–2019) through remote sensing, geographic information system (GIS) analysis, and field-based surveys. This research provides an ideal opportunity to examine a natural erosion event, referred to as cannibalization on a coastal barrier and its dune system. Since the beginning of this century, significant erosion has been visible on the coastal barrier, with the erosion eventually leading to a breaching event in the winter of 2008/2009. Over the study period, analysis has shown that the vegetated dunes decreased by more than 60 percent, the width of the breached area reached a maximum width of over 1 km and a change in orientation and appearance on the coastal barrier has been quantified. The analysis identifies a growing drift-aligned zone, contrasted with a reduction in the stable swash-aligned zone. Significantly, the point between these zones (i.e., the hinge point) has been shown to have moved by more than 1 km also. The migration of this hinge point and cannibalization of the dunes are illustrated. Finally, the potential mechanism for beach healing is identified, utilizing the rich datasets collected during the study, thus providing an insight into the long-term behavior of a dynamic coastal barrier system undergoing naturally driven cannibalization.

Keywords: coastal dunes; barrier dynamics; overwash; coastal erosion; dune cannibalization

1. Introduction

Coastal barriers are narrow, elevated ridges of sand or gravel, often parallel to the shore. The dune systems on sand based coastal barriers play an important role in the coastal environment as they protect the exposed coastlines from open-ocean forcings [1–3]. This also makes them one of the most vulnerable coastal systems. For example, breaching events of the dune system can occur under extreme weather events and after several erosion cycles. Breaching events around the world have been widely reported by [4–7]. Often, these breaching events include human intervention in order to re-establish the dune systems, such as mechanical breach closure, beach scraping, and sediment bypassing systems. Understanding the different processes involved after a breaching event occurred is fundamental for an adequate management of coastal barriers and their dune system.

The spatial and temporal patterns of open-ocean forcings are responsible for the changes prevalent on coastal barriers and dune systems. These include: (1) littoral processes (i.e., the movement of sediment along the coastal region by currents primarily induced by waves and tides); (2) aeolian processes (i.e., transporting wind regime) [8–11]; and (3) other sedimentary factors (e.g., sediment budget, vegetation, backshore budget, and barrier dynamics).

The magnitude and direction of these processes have an influence on the morphology, which could lead to erosion and eventually the breaching of the dune systems on coastal barriers [12]. It is therefore essential to examine the factors that control the spatial and temporal patterns of erosion on dunes and their accretion. These control factors are relatively well understood on small timescales [13–16], whereas their influence and changes in medium and long-term timescale are less clear. As such, it is critical to develop a better understanding of the morphodynamics and hydrodynamics of dune systems for the prediction of the long-term behavior of such systems.

The aim of this study is to understand the drivers and consequences of a natural cannibalizing process. [17] stated that the breaching of a coastal barrier and its dunes can be divided into two different processes: (1) macro-cannibalization and (2) micro-cannibalization. These processes of cannibalization include the change in orientation and the evolution of alignment on coastal barriers. Micro-cannibalization is usually the result of along-shore sediment transport, where the erosion is localized in smaller sub-cells and often results in a breaching event. Contrary to the sub-cell erosion rates of micro-cannibalization, macro-cannibalization is experienced over the entire length of a coastal barrier and its dunes. This is mainly caused by a change in the sediment supply regime and often indicates a change in orientation (i.e., from drift-aligned zone to a swash-aligned zone or vice versa). Both processes have been identified as mechanisms during the evolution of the coastal barrier in Rossbeigh.

This study evaluates the evolution of the dune system on the Rossbeigh coastal barrier over a period of 19 years (2001–2019) with a particular emphasis on the processes prevalent on the coastal barrier and its surrounding coastal cell. The study is a rare opportunity to investigate a natural breaching event as it includes significant datasets from both before and after the breaching event. This enables the authors to undertake an in-depth evolutionary analysis of the system. A collation of the storm events occurring in the study area since 1995 provides context to the drivers that led to the changes in the coastal cell that includes Rossbeigh. The evolution is documented through the analysis of both remotely sensed data and field-based data utilizing GIS tools.

2. Study Area

The Rossbeigh coastal barrier is situated in Dingle Bay, County Kerry, Ireland. It is in the south-western part of Ireland, with longitude around: 9° 58′ 0″ W and latitude around: 52° 05′ 00″ N. The bay is around 18 km wide and 42 km long, and consists of three mid-bay barriers: Rossbeigh, Inch, and Cromane (Figure 1). There is a fast-flowing channel between Inch and Rossbeigh of approximately 2 km in width, where tidal currents can reach over 1 m/s at peak floods, and an expansive ebb tidal delta at the northern (distal) point of Rossbeigh [18]. The tidal inlet between Inch and Rossbeigh acts as an important driver of sediment transport. Dingle Bay itself is characterized as mesotidal, with a spring tidal range of approximately 3.2 m. The mean significant wave height (Hs) is 2.8 m, with an average wave period of Tp = 7 s, taken over a 50-year period of storm data analysis from the M3 wave buoy, located 100 km from the study area (51.2166°N, 10.5500°W) (Sala, 2010). Because Dingle Bay is relatively narrow and bounded by two rocky headlands, the morphology is only affected by a small band of wave forcing, ranging from 225° to 315°. Wind forcing incidents in the bay are predominantly south westerly, as documented in [19]. [19] also classified the coastal cell as a mixed wave/tide dominated to tide dominated system based on [20] and as a self-contained system (i.e., the sediment transport is conserved within the bay itself). Sediment grain size analysis on Inch and Rossbeigh [21,22] indicates that sediments are relatively homogenous with a D_{50} in the range of 260 μm.

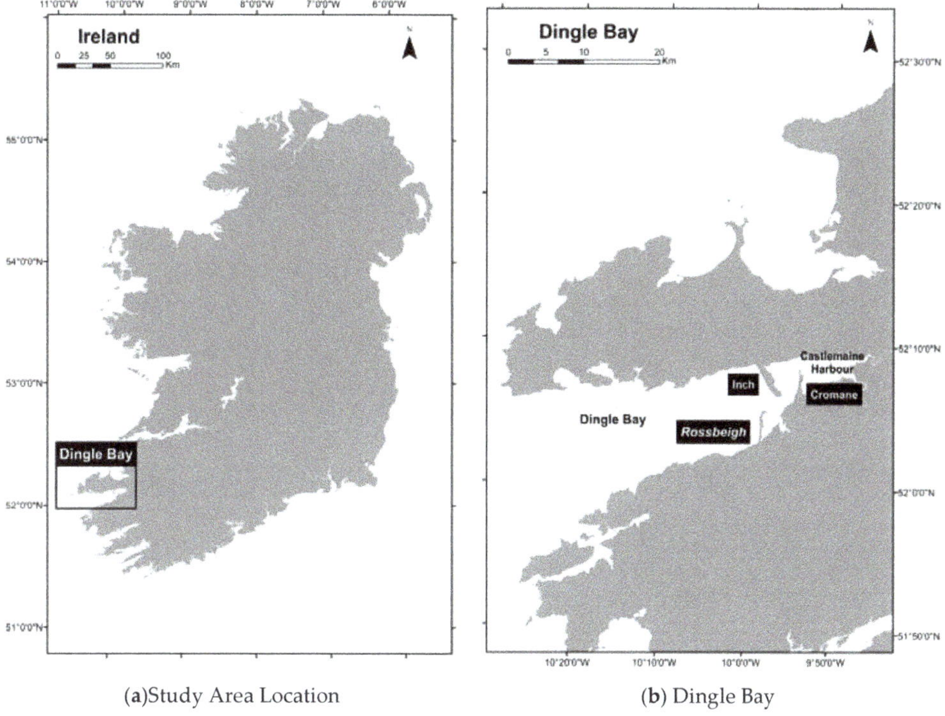

(a)Study Area Location	(b) Dingle Bay

Figure 1. Study Site.

This work focuses exclusively on the Rossbeigh coastal barrier and its dunes. The Rossbeigh coastal barrier extends northwards from the southern shore of Dingle Bay as a relatively stable swash-aligned zone for approximately 3 km (in 2001). The orientation changes and becomes drift-aligned thereafter for approximately 2 km. The drift aligned section of the beach has been shown to be less stable and subject to strong erosion [20,23]. It was [24] that introduced the distinction between these alignments when referring to coastal barriers and beaches and defined their typical properties in terms of evolutionary features i.e., (1) swash-aligned coasts and (2) drift-aligned coasts. The main difference between these two types of coasts is the wave approach perpendicular (1) or an oblique wave approach (2). These distinctions apply to beaches that have no well-developed wave-formed slope on the landward side [25]. The differences in zones are driven by the presence of long-shore drift rates (i.e., drift-aligned zone) or a lack of them (i.e., swash-aligned zone).

The width of the present day Rossbeigh barrier system varies between 100 m and 600 m (Figure 2), and vegetated dunes are present along most of the barrier, apart from the breached area in the drift-aligned zone since the winter of 2008/2009. The barrier itself consists of a sand dune system partially superimposed on a cobble or gravel ridge basement, with the coarser materials acting as an anchor upon which the finer sediments move. The dunes consist of undulating vegetated sand hills on the proximal part (i.e., the section of the coastal barrier attached to the mainland) of the system, with blowouts and low linear foredune ridges fronting the distal part (i.e., the northern section of the coastal barrier on the open coast side). In this case, the proximal part of the coastal barrier is characterized as a swash-aligned zone and the distal part as a drift-aligned zone.

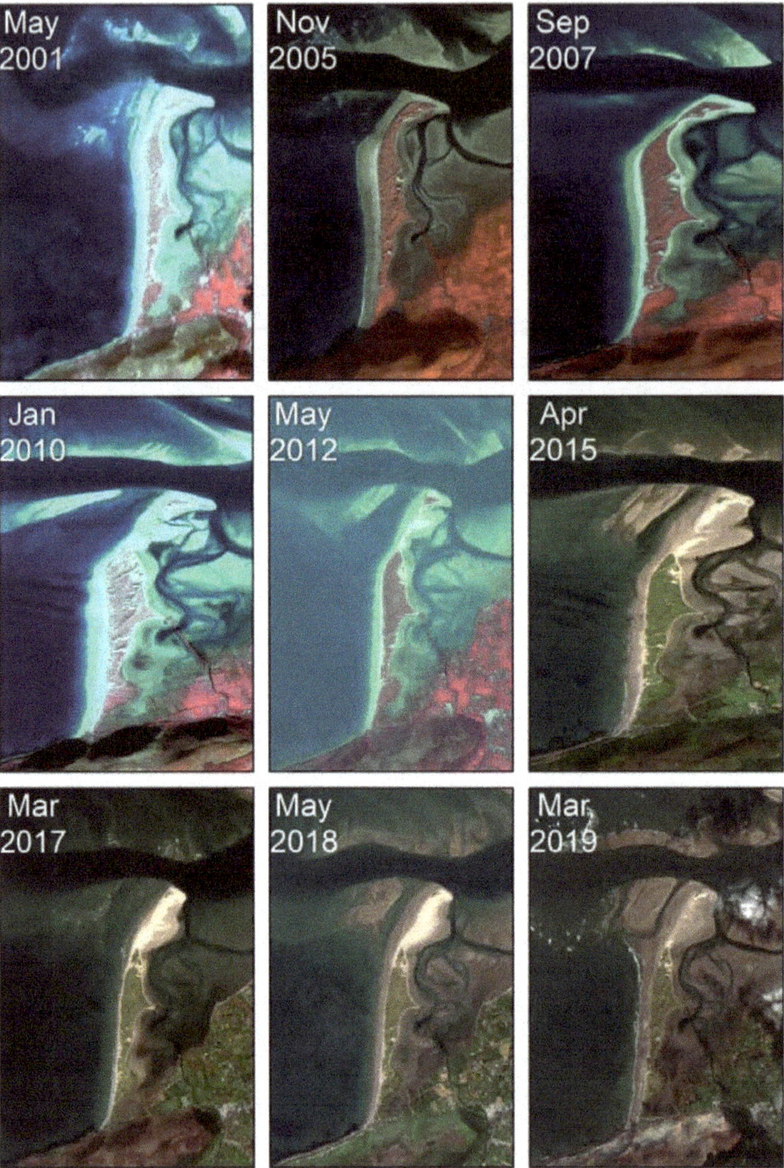

Figure 2. Barrier alignment.

Dune heights in the swash-aligned zone range between 12 m and 17 m above mean sea-level. In the drift-aligned zone, the dune heights decline to values between 5 m and 12 m above mean sea-level. Due to exposure to modally high energy, fully-refracted swell, and the availability of sandy sediment in the drift-aligned zone, the distal part of the coastal barrier is characterized by shallow cross-shore gradients and can be classified as unbarred, dissipative, flat, and featureless, where spilling breakers are dominant [26,27].

In the winter of 2008/2009 a breaching event occurred in the drift-aligned zone of the coastal barrier. This breaching event was the result of continuous erosion from previous years and increased hydraulic forcing by storm surges and waves. This can be described as overwash and is defined as the flow of water and sediment across the crest of the beach/dune that does not directly return to the water body where it originated [28–31]. A breaching event generally takes place in a relatively short time (less than eight hours) [6].

3. Methods

The methodologies utilized in this study include collection and analysis of remotely sensed data, topographic surveying, bathymetric monitoring, and storm event analysis in the study area. The datasets vary in length and temporal resolution, but the majority of analysis focuses on the period since 1995.

3.1. Remote Sensing

Both quantitative and qualitative analysis of the evolution of the coastal barrier in Rossbeigh was carried out with satellite images. The raw images were first imported into an image processing software (i.e., ENVI), which combines advanced image processing and geospatial analysis technology. After initial image processing (e.g., orthorectification and image enhancement), the new data was imported into GIS (i.e., ArcMap 10.6), allowing the visualization of change over time. Table 1 gives an overview of the images used for the analysis. The outline of the coastal barrier in 2001 is used as a reference to show the different changes occurring on the coastal barrier and its surrounding coastal cell.

Table 1. Satellite images used for the analysis of the study area.

Satellite Images Rossbeigh			
Date	**Satellite**	**Sensor**	**Resolution (m)**
21 May 2001	SPOT 4	HRVIR	20
17 November 2005	SPOT 5	HRG	10
09 September 2007	SPOT 5	HRG	10
20 January 2010	SPOT 5	HRG	10
25 May 2012	SPOT 5	HRG	10
17 April 2015	Pleiades	Pleiades 1A	0.5
25 March 2017	Sentinel 2	S2A	10
16 May 2018	Sentinel 2	S2A	10
25 March 2019	Sentinel 2	S2A	10

After analyzing the different images, it was clear that significant erosion occurred on the coastal barrier and its dune system. Thus, it was decided to undertake field-based data collection, including topographical and bathymetric surveys.

3.2. Topographical and Bathymetry Survey

The topographical surveys were completed with three different devices, over different dates. All the equipment used ((1) LEICA 1200; (2) Trimble Ranger 2; and (3) LEICA Viva Rover GS08) use the RTK-GPS method. These devices are global navigation satellite system (GNSS) technologies that track all existing satellite signals, including GPS and GLONASS.

A total of 565 GPS points per survey were collected on the Rossbeigh coastal barrier, in the summer of 2015, 2016, and 2017. These points were imported into ArcMap 10.6 to create and analyze a digital elevation map (DEM). In order to get a more representative model, bathymetric data is collected in the area on the seaward side of Rossbeigh, with special attention given to the ebb tidal bar (EBT) in front of the coastal barrier. This bar is formed by ebb tidal currents depositing material seawards of Rossbeigh. It plays an important role in circulation and sediment transport of the coastal cell.

Bathymetry surveying was undertaken with a personal water craft (PWC) and using a LEICA Viva Rover GS08 GNSS system, in combination with a SonarM8 single beam echo sounder. The SonarM8 has a four-degree beam-width, works on a frequency of 200 kHz, and has a depth range between 0.3 m to 75 m.

Good weather conditions are essential for fieldwork, as wind speed, wind direction, and swell impact data quality. Bathymetric surveys began during high tide, to cover a wider range. To minimize the error sources during surveying, bathymetry was only collected during calm conditions Hs < 0.5 m. This was based on field trials of single beam echo sounder PWC craft surveys recorded by [20] and [32]. Additionally, the accuracy of the points was dependent on the speed of the PWC. Therefore, a speed limit of 8–10 km/h was maintained during surveys.

Bathymetric surveys were undertaken in August 2015 and May 2019 during suitable conditions. The results of these surveys were then compared with a previous bathymetric survey carried out in August 2013 [20].

3.3. Storm Events

In order to define a storm that affects the morphology in the surrounding area of the Rossbeigh coastal barrier, it was decided to use several thresholds for different parameters. Hourly data was available from a nearby weather station: Valentia Observatory (51° 56′ 23″ N, 10° 14′40″ W) (Met Eireann, 2019), providing data for: (1) wind speed, (2) wind direction, and (3) atmospheric pressure. In addition, historical tidal data was generated using a toolbox available in MIKE 21. In situ sea level heights were collected over a period of one month (March 2006–April 2006) using a tidal gauge. The data was collected on Cape Clear, Ireland, located 75 km south of the study area. The data was then extrapolated using the MIKE 21 Global Tide Model in order to get sea level height from 1995 until 2019.

Firstly, the threshold for wind speed was set at 20 knots. On the Beaufort scale, this corresponds to "fresh breeze", with a wave height of 2–3 m and crested wavelets that form on inland waters. As such, 20 knots was considered a threshold that produces waves that will impact the morphology in the coastal cell of Dingle Bay. Secondly, the threshold for wind direction was set at 225–315 degrees. This interval was chosen because Dingle Bay is only affected by storms from this angle due to the topography of the bay, as discussed earlier. Furthermore, the threshold for atmospheric pressure was 990 mbar, as storm events are associated with low pressure areas. Lastly, the threshold for water level was set for water levels above mean sea level for a given period. In addition, a winter/summer split was chosen in order to see if the recovering period, during the summer, was sufficient enough (winter from 1 October–31 March and summer from 1 April–30 September).

In summary, if the wind speed was, for at least one hour, above 20 knots and came from a south-western direction, in combination with an atmospheric pressure below 990 mbar and tides above mean sea level, it was considered a significant storm event. The total amount of storm events and the durations of single storm events occurring in the study area had an impact on the morphology of the coastal cell and drove both the nature and magnitude of the changes incident on the Rossbeigh coastal barrier.

4. Results

The dune vegetation lines were used for the analysis of the changes prevalent on the coastal barrier in Rossbeigh. The change of the vegetation lines prior to and after the breaching event is highlighted in Figure 3. These reference lines were used as a reference to examine the extent of the breached area and to undertake quantitative analysis of: (1) the extent of the vegetated dunes on Rossbeigh, (2) the width of the breached area, and (3) the quantity of the erosion on the coastal barrier.

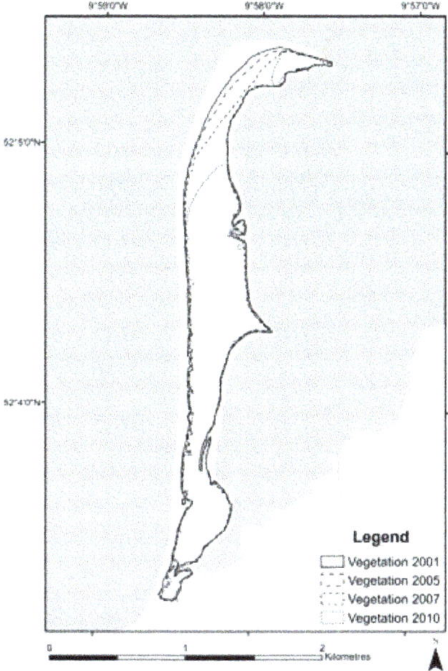

Figure 3. Vegetation alignment of Rossbeigh.

From 2005 onwards, it was clear that erosion was taking place in the drift-aligned zone of the coastal barrier, as the extent of the vegetation was getting narrower towards 2007. In the winter of 2008/2009, overwash led to the breach of the dunes in the drift-aligned zone due to continuous erosion of the previous years. The post-breach vegetation lines are highlighted in Figure 3, revealing the extent of erosion. This was mainly prevalent in the drift-aligned zone in 2010 and 2012, and from 2015 onwards, also in the swash-aligned zone. The extent of the vegetation located on the coastal barrier was observed growing for the first time between 2018 and 2019 since the erosion cycle started in 2001.

A quantitative summary of the depletion of the area of vegetation on the coastal barrier and its dune system can be seen in Table 2. Initially, the total area of vegetated dunes was 1.18 km^2 in 2001. Before the breach in 2008/2009, erosion already had led to a decrease of around 17% to a total area of 1 km^2. After 18 years, a total decrease of 43.2% of the vegetated dunes on the Rossbeigh coastal barrier was apparent.

Table 2. Change in dune area on the coastal barrier from 2001–2019.

Year	Vegetated Area (km^2)	Percentage Change (%)
2001	1.18	
2005	1.10	−7.0
2007	0.99	−10.4
2010	0.89	−12.1
2012	0.78	−13.6
2015	0.71	−10.4
2017	0.68	−3.3
2018	0.65	−5.1
2019	0.67	+2.27
Total	**−0.51**	**−43.2**

Table 3 shows the width of the breached area. Over nine years (2009–2018), the width of the breached area tripled from 325 m in 2009 to a maximum of 1025 m in 2018, a corresponding gradual decrease of the erosion on the coastal barrier is evident also.

Table 3. Width breached area Rossbeigh.

Year	Length (m)
2009 (Sala, 2010)	325
2010	670
2012	845
2013 (O'Shea, 2013)	900
2015	985
2017	1015
2018	1025
2019	1018

As stated earlier, the average dune height ranged between 8–12 m. This is, however, a simplification of the reality, as the height of the dunes was lower in the vicinity of the breach and higher in the swash-aligned zone. Indeed, the dune height exceeded the height of 15 m in some parts and decreased to around 5 m close to the breached area. Hence, two different calculations were undertaken for the volumetric change on the coastal barrier (i.e., dune height of 8 m and a dune height of 12 m). An overview of the volumetric changes since 2001 can be seen in Table 4.

Table 4. Volumetric change Rossbeigh.

Year	Volume Vegetated Dunes-8 m (m^3)	Volume Vegetated Dunes-12 m (m^3)	Volume Change 8 m (m^3)	Volume Change 12 m (m^3)
2001	9,448,480	14,172,720	-	-
2005	8,830,320	13,245,480	−618,160	−927,240
2007	7,998,080	11,997,120	−832,240	−1,248,360
2010	7,137,400	10,706,100	−860,680	−1,291,020
2012	6,282,792	9,424,188	−854,608	−1,281,912
2015	5,660,376	8,490,564	−622,416	−933,624
2017	5,374,720	8,062,080	−285,656	−428,484
2018	5,240,744	7,861,116	−266,088	−99,132
2019	5,362,672	8,044,008	+121,928	+182,892
Total	-	-	**4,084,144**	**6,126,216**

In 2001, the total volume of sediment available in the dunes of the coastal barrier was between 9.4–14.1 million m^3. 18 years later, a total volume of between 4–6.1 million m^3 sediment was eroded from Rossbeigh. At its lowest point in 2018, the total volume of sediment in the dunes was between 5.2–7.8 million m^3.

It is clear from these results (Figures 2 and 3, and Table 2) that the vegetated dune area significantly decreased when the pre-breached (i.e., 2001–2007) situation is compared with the post-breached situation (i.e., 2010–2019).

Apart from the continuous erosion on the coastal barrier, Rossbeigh is also subject to a change in orientation manifesting as macro-cannibalization. This is mainly seen on the drift-aligned zone, whilst the swash-aligned zone is more or less in a state of dynamic equilibrium. The change of orientation can be seen in Figure 4 for 2001– 2018 (no significant change is visible between 2018 and 2019, so 2019 is not included).

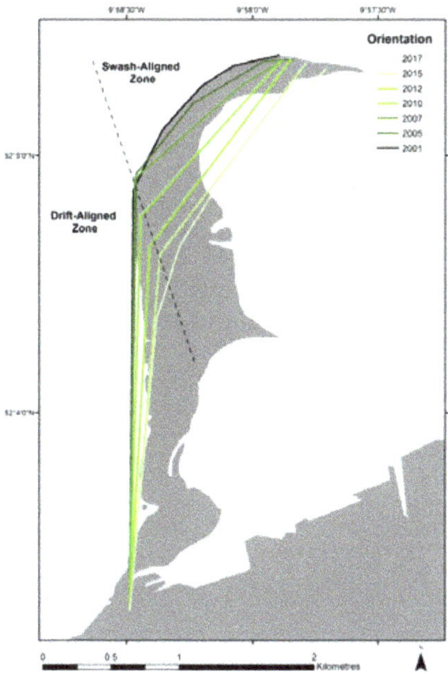

Figure 4. Progression from pre- to post-breach orientation of zones on Rossbeigh.

In addition to the alteration in orientation, the point where the swash-aligned zone shifts to the drift-aligned zone is also moving. The point between those two zones is described as the "Hinge Point" and is defined by the visual change in orientation from the satellite images. Comparison of the Hinge Point location from 2001 – 2019 are compared in Figure 5. It is clear that the Hinge Point was migrating in a northern direction before the breach. However, after the breaching event in 2008, the Hinge Point started moving to the south-east. By May 2018, the Hinge Point moved almost 450 m south and around 80 m east, compared to July 2001. From 2017, there is no clear point visible between the two zones and a "Transition Zone" between the drift-aligned and swash-aligned zone has emerged. The EBT, located on the seaward side of the coastal barrier, is increasing in size and moving towards the coastal barrier. The EBT reached the coastal barrier where it was located in 2001. However, because of the change in orientation, the EBT is still some distance away from the current position of the coastal barrier. There is a difference between the northern and southern part of the bar, with more stable conditions in the north. Moreover, the depth of the seabed was increasing in the north and decreasing in the south, highlighting sediment movement in the channel towards the tidal inlet in Dingle Bay.

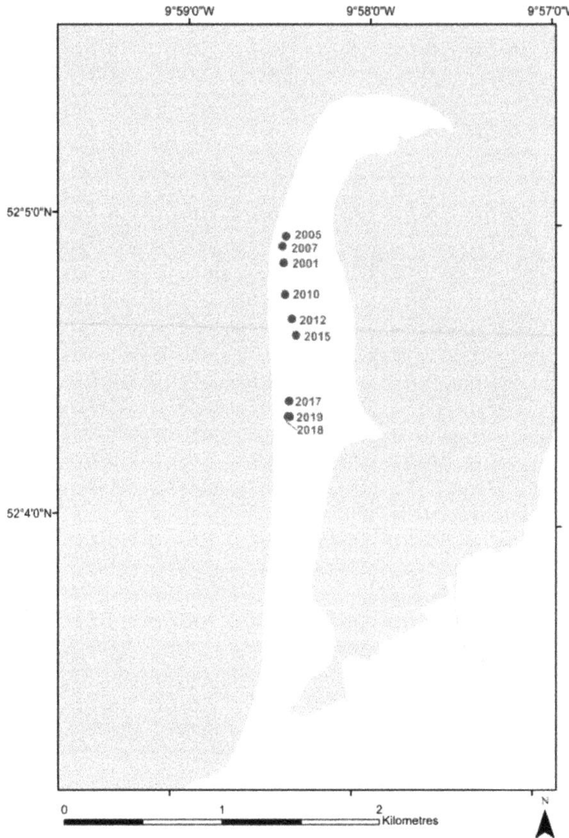

Figure 5. Hinge point migration.

A cross section analysis of the littoral zone, Figure 6 highlights the change of the EBT over time. The analysis was carried out using the different bathymetric surveys. The channel between the EBT and the coastal barrier was migrating, as can be seen from the profiles from the successive bathymetric surveys. Section A-A, at the northern end of this channel, shows that the offshore tidal bar has been migrating towards the drift-aligned zone. The deepest part of the channel between the shore and bar migrated eastwards between 2013 and 2015. With the shoreward progression of the EBT, tidal currents are forced through a narrowing channel (i.e., the channel gets smaller over time). Examining the trend further south, Sections B-B and C-C, it is evident that the channel has been filling in rapidly compared to A-A. This was likely due to the larger wetted perimeter and the larger distance between the EBT and the drift-aligned zone in these locations. The profile is flatter, and the tidal current has a larger if shallower cross-sectional flow area to pass through. By comparison the more northerly section A-A is deeper but narrower and more dynamic.

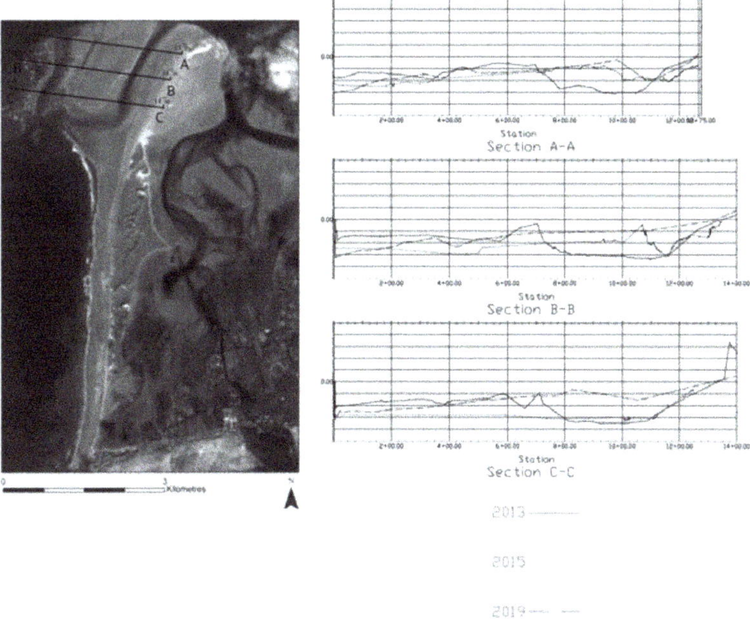

Figure 6. Survey cross section comparison.

The adjustment of the coastline can arise during a single storm event or during a series of closely spaced storm events. Extreme weather events occurring in the study area in the late 1990s' and early 2000s' were likely the cause of the erosion rates that occurred on the coastal barrier of Rossbeigh. Figure 7 highlights the number of storm events from winter 1994/1995 to the summer of 2019, based on the thresholds discussed earlier. These storms were analyzed more specifically in relation to their potential impact on morphodynamics on Rossbeigh by calculating the average duration of an individual storm event, Figure 8, and the total duration of all storm events in a given season, Figure 9.

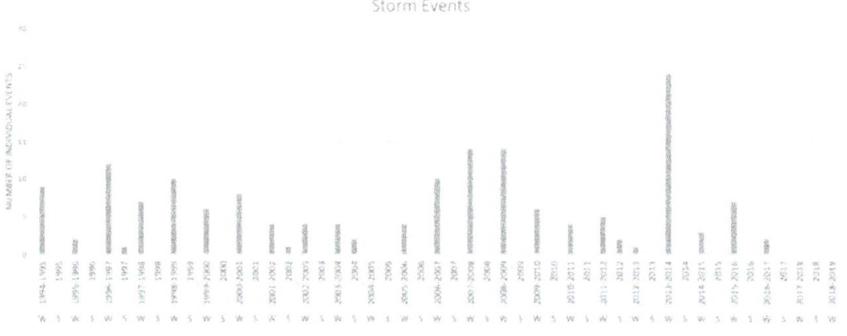

Figure 7. Number of storm events in the preceding 25 years incident on Dingle Bay.

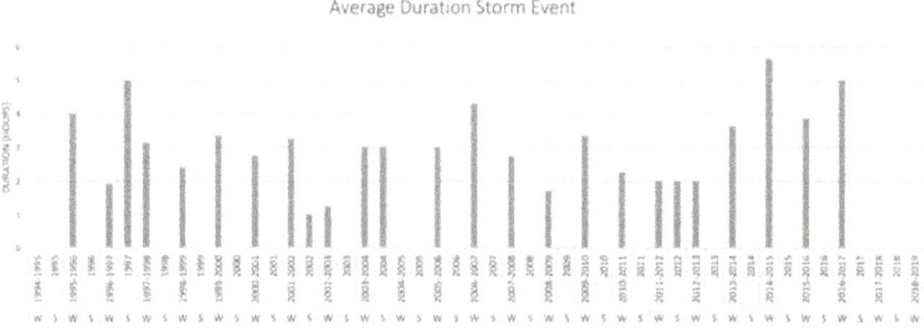

Figure 8. Mean storm durations in the preceding 25 years incident on Dingle Bay.

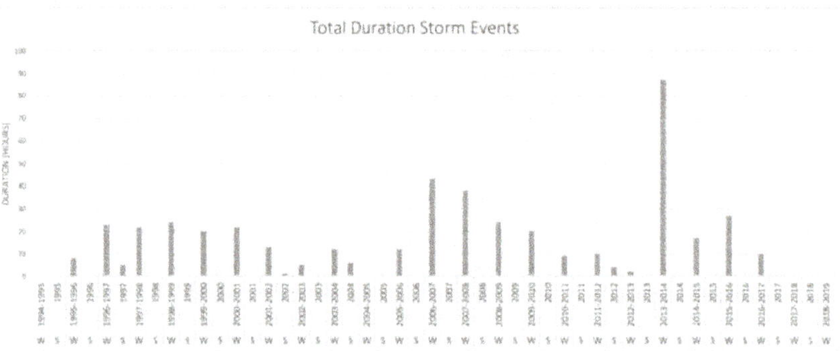

Figure 9. Total duration of storm events in the preceding 25 years incident on Dingle Bay.

It is clear that, prior to 2001, a significant amount of winter storms occurred. These events were probably the trigger that led to the destabilization of the morphology in the study area and the start of the erosion events. Before the breaching event, there was another peak in storm events, arguing that they are eventually the cause of the breach and overwash. Afterwards, another peak in storm events is visible in the winter of 2013–2014.

5. Discussion

Over the last 19 years (2001–2019), significant erosion has taken place on the Rossbeigh coastal barrier and its dunes. The erosion cycle that started in the early 2000s led to the current breaching event, which was caused by overwash, due to forcings from the western seaward side [18]. Atypical from other tidal inlets that are formed through breaching, the inlet that formed on the Rossbeigh coastal barrier is still open 10 years after the breaching event [33], and indicates that tidal inlets formed due to breaching should be able to heal over the years. It is clear, however, that until 2018, this system still had to attain some form of post-breaching equilibrium, as little or no seasonal recovery was visible during the time of observation and cannibalization was ongoing for a long period after the breaching event. The first signs of recovery are visible in 2019. For the first time since the breaching event, the vegetated area of the dunes is growing.

The growth of the EBT and the movement of that bar towards the coastal barrier suggest that the sediment eroded over the years during the cannibalization process will fill the channel fronting the Rossbeigh coastal barrier. We argue that this will eventually lead to the healing of the breach and will result in the possible restoration of the dune system prevalent on the coastal barrier prior to the erosion cycle characterized by micro- and macro-cannibalization that dominated since beginning of this century. Prior to 2017, there was no documented evidence that such a phase had commenced; the analysis of both remotely sensed data and the bathymetric survey shows that there is a clear migration trend of the EBT. The present study has confirmed that the EBT was moving eastwards towards the drift-aligned zone, from as early as April 2015. The analysis of this study also shows that the bar is in the position where the Rossbeigh coastal barrier was located in 2001.

Since the summer of 2017, marram grass has started to colonize areas in the breached area. This suggests that more stable conditions around the coastal cell of Rossbeigh favor the healing of the breach. Although, the erosion and the breached area was still increasing in 2018, signs of recovery were finally visible from 2019. This is likely linked with the growth of the EBT, which now covers the whole area in front of the breached area. This means that the impact of the open ocean forcings has decreased, and larger waves are not reaching this vulnerable section of the coastal barrier anymore. In addition, the analysis of the storms indicates a milder climate incident on the barrier beach compared to that of the large storm events that were previously occurring in the area. The conditions were more favorable in the last two years, 2017 and 2018, allowing breach healing now. If the EBT continues to grow, we argue that breach healing will occur in five years, with dune establishment to commence thereafter.

The change in orientation and the movement of the hinge point in a south-eastern direction shows that the swash-aligned zone is still decreasing. It is obvious from the figures highlighted in this study that there is a growing drift-aligned zone at the expense of the more stable swash-aligned zone. As the drift-aligned zone continues to grow, it is expected that macro-cannibalization will be ongoing. The decrease of the swash-aligned zone has left the growing drift-aligned zone and transition zone of the coastal barrier more vulnerable to wave action, as this part is not protected by the EBT. As such, the foredunes prevalent near the hinge point (i.e., the foredunes in the vicinity of the breached area) are subject to increased erosion and are particularly vulnerable during storm events.

This was especially visible between 2017 and 2018, where the swash aligned zone was clearly eroding as well. However, there is no distinct difference anymore between the swash-aligned zone and the drift-aligned zone. The transition zone between the swash-aligned zone and the drift-aligned zone is growing, which leads to a decrease in the swash-aligned zone. Hence, it can be argued that micro-cannibalization will eventually end, which will lead to the regeneration of the Rossbeigh coastal barrier and its dune systems. However, it is still unclear how macro-cannibalization will evolve over time, making it a highly dynamic and unpredictable system.

Storm Events

The morphological response of the beach profile is slower compared to the faster variation of the hydrodynamic forcing. As such, recovery, or the lack thereof, can extend far beyond a single season [34]. Changes in the morphology caused by storm events are often reversible if the system repairs itself during normal conditions. The changes in Rossbeigh, however, could be irreversible if the new morphology changes the sediment transport regime and the hydrodynamics of the system to such an extent that recovery is impossible within an immediate timeframe (i.e., less than decadal).

Kandrot, S. et al. [35] argues that storm events occurring in the study area are responsible for the delay and/or prevention of such beach healing. Storm events occur frequently in the southwest of Ireland. Studies [3,26,36] have shown that storm events can have a significant impact on the sediment budgets of barrier systems. In this regard, the morphodynamics within such systems are dependent on the seasonal behavior of the tidal inlet, which is mainly affected by storm events.

The absence of big storm events in the last seasons and the recovery of the dune system in 2019 support this theory. The analysis of the channel between EBT and drift aligned shore suggests that such storm events may force accelerated migration of the EBT. This shoreward migration has been shown as a potential mechanism of breach healing. It is proposed that storm driven shoreward sediment transport could close the channel. The rate of closure of the channel is shown to have accelerated between 2013 and 2015, which corresponds to the frequent storm events in 2014. This channel closure acceleration is clearly visible in Section A-A of Figure 7 over that period. During the intervening period since 2015, storms have been more infrequent and channel closure and migration rates have stabilized. It is possible a further increase period of storm activity similar of 2014 is required to provide the final closure of the channel and have the EBT weld to the drift aligned shore. This would lead to a step change in the migration of the EBT, and beach healing could occur rapidly thereafter in the absence of regular scouring that the tidal channel is currently providing. The observed accretion rate would increase and allow aeolian transport and dune vegetation restorative processes to initiate dune regeneration. In this regard, continuous monitoring of the system is important in order to highlight the change over time of a natural breaching event.

6. Conclusions

It is clear from the examination of 19 years of remotely sensed data combined with field measurements that the Rossbeigh coastal barrier was in a post-breaching phase for the majority of the monitoring period. However, analysis of the dataset signifies that the system is capable of reaching a post-breaching equilibrium. This work identifies tangible processes that are occurring (e.g., movement of the EBT towards the coastal barrier and colonization of marram crass in the breached area) that indicate regrowth and breach healing is possible under the right circumstances (e.g., the protection of open ocean forcings by the EBT). The role of storm events within the study area suggests that while the system has not yet been able to reach a full beach-healing phase, a mechanism for such healing has been identified.

It is proposed that the regenerative gains observed during calm conditions are temporary and that further storm driven dune erosion is likely until such time as the EBT welds to the drift aligned shore. The storm events may play an important role in this process, as they mobilize large volumes of sediment. It is proposed that a step change is required to overcome the established tidal current regime that is preventing full migration and welding of the EBT to the drift aligned shore. A significant storm or series of storm events would provide the step change required for the system to enter a new cycle of growth and breach healing. It is vital therefore to map and monitor the different changes that occur on such systems in order to develop a clear understanding of the long-term behavior of coastal barriers and how the evolutionary phases transition from erosive to accretive.

Further Research

Regular data collection and analysis makes it possible to make predictions on the evolution of the coastal barrier in Rossbeigh and its surrounding coastal cell. In this regard, it is suggested to do follow-up surveys in order to get datasets with high temporal and spatial accuracy. This will lead to a more detailed analysis and to more accurate predictions.

It is necessary to achieve a complete understanding of the coastal processes in the study area before any human intervention is undertaken. This confirms the necessity of this research and future studies in Rossbeigh and its surrounding coastal cell. It is suggested to carry out a grain size trend analysis which associates grain sizes with sediment transport pathways on the Rossbeigh coastal barrier. This study would be similar in nature to the previous exercise by [18] but on a wider geographic scale over several seasons. Lastly, other results highlighted in this research (e.g., location of hinge points) were not analyzed in detail. A numerical analysis of zone orientation and hinge point location combined, including a curve fitting exercise to the shoreline, could provide further insight. This opens up possibilities with establishing mathematical relationships for the evolution such as second derivatives. Furthermore, the historical analysis undertaken on Rossbeigh and Inch in previous studies could be reanalyzed with such a mathematical approach. If such work is to be carried out the information produced in this study will form an essential baseline.

Author Contributions: Conceptualization, S.N.; methodology, S.N. and M.O.; software, S.N. and M.O.; investigation, S.N. and M.O.; writing—original draft preparation, S.N.; writing—review and editing, J.M.; visualization, S.N.; and supervision, J.M. All authors have read and agreed to the published version of the manuscript.

J. Mar. Sci. Eng. **2020**, *8*, 421

Funding: This research received no external funding.

Conflicts of Interest: The authors declare no conflict of interest.

References

1. Saye, S.E.; Pye, K. Implications of sea level rise for coastal dune habitat conservation in Wales, UK. *J. Coast. Conserv.* **2007**, *11*, 31–52. [CrossRef]
2. Blott, S.J.; Pye, K. Particle shape: A review and new methods of characterization and classification. *Sedimentology* **2008**, *55*, 31–63. [CrossRef]
3. O'Shea, M.; Murphy, J.; Sala, P. Monitoring the morphodynamic behaviour of a breached barrier beach system and its impacts on an estuarine system. In Proceedings of the 2011 IEEE OCEANS, Santander, Spain, 6–9 June 2011; pp. 1–8.
4. Chadwick, A.; Karunarathna, H.; Gehrels, W.; Massey, A.; O'brien, D.; Dales, D. A new analysis of the Slapton barrier beach system, UK. *Proc. ICE-Marit. Eng.* **2005**, *158*, 147–161. [CrossRef]
5. Freeman, C.W.; Bernstein, D.J.; Mitasova, H. Rapid response 3D survey techniques for seamless topo/bathy modelling: 2003 Hatteras Breach, North Carolina. *Shore Beach* **2004**, *72*, 25–29.
6. Gracia, V.; García, M.; Grifoll, M.; Sánchez-Arcilla, A. Breaching of a barrier beach under extreme events. The role of morphodynamic simulations. *J. Coast. Res.* **2013**, *65*, 951–956.
7. Terchunian, A.; Merkert, C. Little Pikes Inlet, Westhampton, New York. *J. Coast. Res.* **1995**, *11*, 697–703.
8. Tsoar, H.; Levin, N.; Porat, N.; Maia, L.P.; Herrmann, H.J.; Tatumi, S.H.; Claudino-Sales, V. The effect of climate change on the mobility and stability of coastal sand dunes in Ceará State (NE Brazil). *Quat. Res.* **2009**, *71*, 217–226. [CrossRef]
9. Nickling, W.G.; Davidson-Arnott, R.G. *Aeolian Sediment Transport on Beached and Coastal Dunes. Proceedings Symposium on Coastal Sand Dunes*; NRC: Ottawa, ON, Canada, 1990; pp. 1–35.
10. Ritchie, W.; Penland, S. Rapid dune changes associated with overwash processes on the deltaic coast of South Louisiana. *Mar. Geol.* **1988**, *81*, 97–122. [CrossRef]
11. Arens, S.M. Rates of aeolian transport on a beach in a temperate humid climate. *Geomorphology* **1996**, *17*, 3–18. [CrossRef]
12. O'Shea, M.; Murphy, J. Investigating the Hydrodynamics of a Breached Barrier Beach. *EGU Gen. Assem. Conf. Abstr.* **2011**, *14*, 43.
13. Edelman, T. Dune Erosion During Storm Conditions. In *Proceedings of the 11th Conference on Coastal Engineering, London, UK, September 1968*; American Society of Civil Engineers: Reston, VA, USA; Volume 2, pp. 719–722. [CrossRef]
14. Vellinga, P. Beach and dune erosion during storm surges. *Coast. Eng.* **1982**, *6*, 361–387. [CrossRef]
15. van de Graaff, J. Dune erosion during a storm surge. *Coast. Eng.* **1977**, *1*, 99–134. [CrossRef]
16. Steetzel Henk, J. Cross-Shore Transport during Storm Surges. *Coast. Eng.* **1990**, *1991*, 1922–1934. [CrossRef]
17. Orford, J.D.; Carter, R.W.G.; Jennings, S.C. Control Domains and Morphological Phases in Gravel-Dominated Coastal Barriers of Nova Scotia. *J. Coast. Res.* **1996**, *12*, 589–604.
18. Gault, J.; O'Hagan, A.M.; Cummins, V.; Murphy, J.; Vial, T. Erosion management in Inch beach, South West Ireland. *Ocean Coast. Manag.* **2011**, *54*, 930–942. [CrossRef]
19. O'Shea, M. Monitoring and Modelling the Morphodynamic Evolution of a Breached Barrier Beach System. Ph.D. Thesis, University College Cork, Cork, Ireland, 2013.
20. FitzGerald, D.; Kraus, N.; Hands, E.B. *Natural Mechanisms of Sediment Bypassing at Tidal Inlets*; US Army Corps of Engineers: Vicksburg, MS, USA, 2000.
21. O'Shea, M.; Murphy, J. The Validation of a New GSTA Case in a Dynamic Coastal Environment Using Morphodynamic Modelling and Bathymetric Monitoring. *J. Mar. Sci. Eng.* **2016**, *4*, 27. [CrossRef]
22. Wintle, A.G.; Clarke, M.L.; Musson, F.M.; Orford, J.D.; Devoy, R.J.N. Luminescence dating of recent dunes on Inch Spit, Dingle Bay, southwest Ireland. *Holocene* **1998**, *8*, 331–339. [CrossRef]
23. Sala, P. Morphodynamic Evolution of a Tidal Inlet Mid-Bay Barrier System. Master's Thesis, University College Cork, Cork, Ireland, 2010; 226p.
24. Davies, J.L. *Geographical Variation in Coastal Development*, 2nd ed.; Longman: New York, NY, USA, 1980.
25. Haslett, S. *Coastal Systems*; Routledge: London, UK, 2008.

26. Williams, J.J.; Esteves, L.S.; Rochford, L.A. Modelling storm responses on a high-energy coastline with XBeach. *Modeling Earth Syst. Environ.* **2015**, *1*, 3. [CrossRef]
27. Masselink, G.; Short, D.A. The Effect of Tide Range on Beach Morphodynamics and Morphology: A Conceptual Beach Model. *J. Coast. Res.* **1993**, *9*, 785–800.
28. Kraus, N.C.; Wamsley, T.V. Coastal Barrier Breaching. Part 1. Overview of Breaching Processes, Coastal and Hydraulics Engineering Technical Note ERDC/CHL CHETN-IV-56, US Army Corps of Engineers, Vicksburg, MS, USA, 2003. Available online: http://chl.wes.army.mil/library/publi-cations/chetn/pdf/chetn-iv-56.pdf (accessed on 28 March 2020).
29. Donnelly, C.; Kraus, N.; Larson, M. State of Knowledge on Measurement and Modeling of Coastal Overwash. *J. Coast. Res.* **2006**, 965–991. [CrossRef]
30. Matias, A.; Williams, J.J.; Masselink, G.; Ferreira, Ó. Overwash threshold for gravel barriers. *Coast. Eng.* **2012**, *63*, 48–61. [CrossRef]
31. McCall, R.T.; Van Thiel de Vries, J.S.M.; Plant, N.G.; Van Dongeren, A.R.; Roelvink, J.A.; Thompson, D.M.; Reniers, A.J.H.M. Two-dimensional time dependent hurricane overwash and erosion modeling at Santa Rosa Island. *Coast. Eng.* **2010**, *57*, 668–683. [CrossRef]
32. MacMahan, J. Hydrographic surveying from personal watercraft. *J. Surv. Eng.* **2001**, *127*, 12–24. [CrossRef]
33. Davis, R.; Fitzgerald, D. *Beaches and Coasts*; John Wiley and Sons: New York, NY, USA, 2004; 432p.
34. Larson, M.; Erikson, L.; Hanson, H. An analytical model to predict dune erosion due to wave impact. *Coast. Eng.* **2004**, *51*, 675–696. [CrossRef]
35. Kandrot, S.; Farrell, E.; Devoy, R. The morphological response of foredunes at a breached barrier system to winter 2013/2014 storms on the southwest coast of Ireland. *Earth Surf. Process. Landf.* **2016**, *41*, 2123–2136. [CrossRef]
36. Balouin, Y.; Howa, H. Sediment transport pattern at the Barra Nova inlet, south Portugal: A conceptual model. *Geo-Mar. Lett.* **2001**, *21*, 226–235. [CrossRef]

Journal of
Marine Science and Engineering

Article

Cross-Shore Profile Evolution after an Extreme Erosion Event—Palanga, Lithuania

Loreta Kelpšaitė-Rimkienė [1,*], Kevin E. Parnell [2], Rimas Žaromskis [1] and Vitalijus Kondrat [1]

[1]   Marine Research Institute, Klaipėda University, 92294 Klaipėda, Lithuania;
      rimas.zaromskis@cablenet.lt (R.Ž.); vitalijus.kondrat@ku.lt (V.K.)
[2]   Department of Cybernetics, Tallinn University of Technology, 19086 Tallinn, Estonia; kevin.parnell@taltech.ee
*   Correspondence: loreta.kelpsaite-rimkiene@ku.lt

Abstract: We report cross-shore profile evolution at Palanga, eastern Baltic Sea, where short period waves dominate. Cross-shore profile studies began directly after a significant coastal erosion event caused by storm "Anatol", in December of 1999, and continued for a year. Further measurements were undertaken sixteen years later. Cross-shore profile changes were described, and cross-shore transport rates were calculated. A K-means clustering technique was applied to determine sections of the profile with the same development tendencies. Profile evolution was strongly influenced by the depth of closure which is constrained by a moraine layer, and the presence of a groyne. The method used divided the profile into four clusters: the first cluster in the deepest water represents profile evolution limited by the depth of closure, and the second and third are mainly affected by processes induced by wind, wave and water level changes. The most intensive sediment volume changes were observed directly after the coastal erosion event. The largest sand accumulation was in the fourth profile cluster, which includes the upper beach and dunes. Seaward extension of the dune system caused a narrowing of the visible beach, which has led to an increased sand volume (accretion) being misinterpreted as erosion

Keywords: cross-shore profile; sediment transport rates; semi-enclosed sea; sandy coast; coastal erosion; dune development

Citation: Kelpšaitė-Rimkienė, L.; Parnell, K.E.; Žaromskis, R.; Kondrat, V. Cross-Shore Profile Evolution after an Extreme Erosion Event—Palanga, Lithuania. *J. Mar. Sci. Eng.* **2021**, *9*, 38. https://doi.org/10.3390/jmse9010038

Received: 1 December 2020
Accepted: 22 December 2020
Published: 2 January 2021

Publisher's Note: MDPI stays neutral with regard to jurisdictional claims in published maps and institutional affiliations.

1. Introduction

Explaining changes to nearshore coastal profiles remains a challenge for coastal researchers, particularly where there is considerable alongshore sediment transport [1–3]. Changes to the underwater profile are frequently examined under controlled hydrodynamic conditions in wave flumes [4,5], with some studies incorporating the effects of structures [6]. There are numerous examples of beach profile datasets presented in the literature (e.g., [1–3,7–11]), which consider how beaches change in response to variability in wind and wave conditions, advancing understanding with the analysis of each new dataset collected under different conditions. This study examines beach changes at Palanga, Lithuania, a heavily modified beach on a tideless coast with significant longshore sediment transport.

The Baltic Sea's eastern coast is a high-energy (for the Baltic Sea), actively developing coast with fine, highly mobile sediments [12,13]. Sediment transport is generally counter-clockwise along the entire south-eastern coast of the Baltic Proper [14,15] from the Sambian Peninsula to Kolka Cape [16]. Minor variations in the physical nature of the coast and human activities (such as ports and other structures) add complexity to the system's evolution. For example, parts of the underwater slope along the Lithuanian and Latvian shores have boulders, pebbles, and coarse sand. At the same time, in other sections, such as along the Curonian spit, there are fine sands with well-developed bar systems. Only in a few places (the Curonian Spit and a short coastal section to the south-east of Kolka Cape) are there substantial quantities of fine sediment [17]. These coast sections are generally stable or accretionary.

By comparison, other parts of the SE Baltic proper coasts are generally erosive [12,18,19]. The most significant areas of coastal retreat are near the Ports of Klaipeda, Liepaja, Ventspils, and in the vicinity of other structures [18–21]. Shoreline retreat near the Port of Klaipeda has reached 0.75 m/year since 1947 [12,22]. Further north, near the Port of Liepaja, there has been shoreline retreat more than 150 m since 1935, a rate of almost 2.5 m/year, and near the Port of Ventspils, 2 m/year shoreline retreat has been reported [19]. At Palanga (Figure 1), the site of this study, erosion of 0.75 m/year has been reported in the literature [12]. While some areas have been eroding, the well-developed dune systems on the SE Baltic coast have accumulated a large volume of sand [23], in part the result of soft coastal protection such as dune planting [24]. Development of the dune system is well correlated with the amount of vegetation immediately landward of the active beach [25]. Nevertheless, sediments accumulated in the dunes act as protection for the coastal ecosystem, but the ongoing retreat of the shoreline signals a decrease in the amount of sediment transported on the underwater slope.

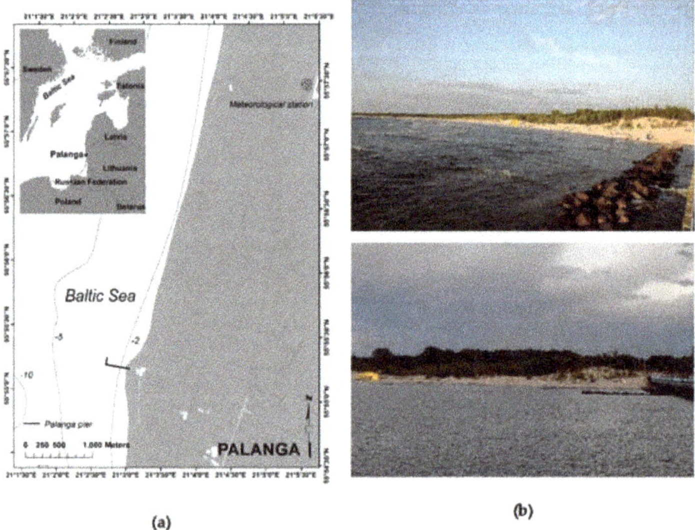

Figure 1. (**a**) Study area and (**b**) the northern side of the pier, with the adjacent groyne system (Photos: September 2018).

Coastal processes on beaches within semi-enclosed or enclosed seas can vary from those on open ocean shores [22]. Swell waves are almost non-existent. Mean wave height on the SE Baltic Seacoast is 0.6 m [26,27]. Highest waves occur from September to January when monthly mean wave height reaches 0.8 m [28], with the overall maximum values 6 m [26]. Waves tend to be a short period (mean wave period about 2.5 s [27,29]), meaning that profile closure depths are limited, and wave refraction occurs close to the beach. Processes that generate significantly elevated water levels [23], such as storm surge [24] and wave set-up [25], which in turn are strongly related to coastal erosion, are highly variable in space and highly localized in their effect. A large proportion of the wave energy flux to the coast occurs on a very few days of the year [26], meaning that the wind direction at the time of major storms is a significant determinant of the coastal change outcomes. Changes in the wind regime, and therefore extreme water levels and wave characteristics [27], may radically alter coastal processes [28]. Few coastal beach profile studies address enclosed-sea environments, where the process-response regimes differ from open-ocean beaches.

Palanga (Figure 1) is the biggest eastern Baltic Sea seaside resort [30]. The first timber pier was built in 1889. It has a solid construction, interrupting the natural hydrodynamics and sediment transport. As a result, by 1892, sand accumulation next to the pier meant that ships were no longer able to moor. The design was adapted to repurpose the pier for recreation and Palanga was developed as a holiday resort. From construction until 1910, the shoreline moved seawards by "500 steps" (~400 m), and through to 1947, it accreted another 100 m [31]. At the end of the 19th century, the coast nearby was relatively flat, but by the middle of the 20th century, the shoreline comprised a dune field 80–100 m wide with 6–8 m high dunes, with a sand volume of about 400 m^3 per linear meter [32]. Over the course of a hundred years, the pier was reconstructed several times, without changing its basic structure. In 1990, a decision was made to build a new concrete pier, permeable to waves and sediments. The new pier was completed in 1995, and demolition of the old impermeable structure started. Coastal erosion was observed soon after the shore perpendicular groynes, which were part of the old design, were removed in 1997 [31,32]. Significant coastal erosion becomes evident after the storm "Anatol" on 4 December 1999.

Storm "Anatol" remains the most damaging storm recorded on the SE Baltic Sea coast [33,34] with predominant west and north-west wind directions, mean wind speed up to 22m/s, and maximum wind gusts reaching 40 m/s [33]. "Anatol" stands out as being unusual for the Lithuanian coast by its trajectory over the Baltic Sea [35] resulting in up to 40 m/s south and south-west winds. Several comparable storms have since crossed Baltic Sea, examples being Ervin (or Gudrun) in 2005 and Kiril in 2007 [36,37] but they were more hazardous on the NE Baltic Sea coast [31,38,39]. During the storm "Anatol", more than 3 million m^3 of sand were eroded from the Lithuanian Baltic Sea coast [34], which at Palanga caused a 35 m reduction in beach width [31] and resulted in the dune base moving inland by 10 m [11,31,32]. The pier was damaged at its landward end and detached from the shore (Figure 2a), and the dune system was destroyed [30]. As a result of the changes, the groyne system (Figure 1b) was rebuilt in May 2000 [32].

(a)   (b)

Figure 2. (a) Palanga pier after the storm "Anatol", December 1999, and (b) in September 2018.

In this paper, we analyze beach underwater profile recovery after the storm "Anatol" and the groyne reconstruction and later profile evolution. While processes associated with erosion of the upper beach and the deposition of sand in the nearshore are relatively well understood, cross-shore sediment transport variations under-recovery conditions have received less attention. The overall aim is to examine cross-shore sediment transport variations under different accretive conditions, beginning with a significantly eroded underwater beach profile, using conventional profile analysis, but incorporating analysis of transport rates and attempting to identify different process sections of profile using cluster analysis.

2. Materials and Methods

The cross-shore profile was surveyed at Palanga (Figure 1) adjacent to the Palanga pier over two time periods. The first set of measurements was undertaken in the period 1999–2000. Measurements began directly after the storm "Anatol" on 7 December 1999 and continued during the year 2000. Wind velocity and direction data for the study period were obtained from the Palanga hydro-meteorological monitoring station, Marine Research Department of the Environmental Protection Agency. Environmental monitoring was performed according to the national monitoring program guidelines prepared by the Lithuanian Ministry of Environment following various legal acts of the European Union [40].

Initially, the field experiment was designed to monitor cross-shore profile changes under different hydrometeorological conditions [11]. Sand composition on the underwater slope is characterized as well-sorted fine sands with a mean diameter of 0.18 mm [41]. Measurements with a weighted line (marked at 10 cm intervals) were made along the southern side of the pier (400 m length) every 2.5 m with zero distance being the pier's seaward-most handrail. Although the water level on the coast oscillates with an amplitude of more than 1.5 m during storm surges [11], the apparent Mean Sea Level (MSL) is considered to be 5 m below the handrails, based on long-term observations. Measurements using the same methodology were undertaken in the stormy seasons of 2016 and 2017, with weekly measurements October–December until such time as there was ice along the coast.

The seaward limit of profile fluctuation over long-term (seasonal or multi-year) time scales is referred to as the "closure depth", denoted by h_c. Based on laboratory and field data, Hallermeier (1978, 1981) developed the first rational approach to determining closure depth [42]. Based on correlations with the Shields parameter, Hallermeier defined a condition for sediment motion resulting from relatively rare wave conditions. Effective significant wave height H_e and effective wave period T_e were based on conditions exceeded only 12 h per year, i.e., 0.14 percent of the time. The resulting approximate equation for the depth of closure was determined to be:

$$h_c = 2.28 H_e - 68.5 \left(\frac{H_e^2}{g T_e^2} \right) \tag{1}$$

$$H_e = \overline{H} + 5.6 \sigma_H \tag{2}$$

$$h_c = 2\overline{H} + 11 \sigma_H \tag{3}$$

where g is gravity, and σ_H is the standard deviation of annual wave heights. Therefore, good approximation to the data is given simply by $h_c = 1.57\ H_e$ [43]. In the case of the Lithuanian coast where $H_e = 4$ m [26], the depth of closure $h_c = 6.3 \pm 0.5$ m [43,44]. However, the storm "Anatol" revealed a layer of hard moraine sediments at ~ -5 m depth, effectively constraining the profile at a level above calculated closure depth [11].

Comparative plots of beach profile evolution over time were constructed for the 1999–2000, 2016, and 2017 year cases. Average profiles for the periods were calculated. The cross-shore transport rates Q(x) were calculated using the methods in [4,5]. The total sediment transport rate (bedload and suspended load) per unit width between any two time periods (interval Δt) is determined from:

$$Q(x_n) = Q(x_{n-1}) - \int_{x_{n-1}}^{x_n} (1-p) \frac{\Delta z_b}{\Delta t} dx \tag{4}$$

where positive values of $Q(x_n)$ (m^2/Δt) represent onshore sediment transport at position n, Δz_b is the difference in the bed elevation between measurement intervals (m), Δt is the time difference between measurements (year), and p is the porosity of the sand, being 0.4 [2,5,9]. We assume no net sediment transport past the run-up limit x_{max} and beyond the depth of closure x_{min}, and sediment transport occurs consistently over the beach profile.

The bulk cross-shore sediment transport Q across the whole profile between any two time periods is determined by integrating the local transported volume along the profile [5]:

$$Q = \Delta t \int_{x_{min}}^{x_{max}} Q(x)dx \qquad (5)$$

between the same closure limits. Q represents the bulk cross-shore sediment transport (m^3 per linear meter) moved either shoreward (positive) or offshore (negative). This measure has been used to categorize the overall beach response as erosive (Q < 0), accretionary (Q > 0) or stable (Q $\approx$ 0). Note that Q is a transport vector and can be either purely negative or positive, or a mixture, and therefore does not integrate to zero unless the onshore and offshore transport magnitudes are equal or both identically zero.

We use K-means clustering to determine clusters of cross-shore profile segments with similar development trends [45]. The K-means algorithm is one of the most popular hierarchical algorithms and uses the minimum sum of squares to assign observations to groups. Such groups of data points are called clusters [46,47]. Observations allocated to the closest cluster, and the distance between an observation and a cluster is calculated from the Euclidean distance between the observation and the cluster center. The objective function of K-means is given as:

$$E = \sum ||X_i - m_i||^2 \qquad (6)$$

where: E is the sum of square error for all objects in the data, X_i is the point in a cluster, and m_i the mean of cluster k_i. The goal of K-means is to minimize the sum of the squares error over all k clusters. The algorithm states that, initially, k points are placed into space represented by objects that need to be clustered as initial group centroids. In the second step, each object is assigned to its closest cluster center. Then, the mean of each cluster is calculated to have a new centroid. These steps are repeated until there is no change in centroids. The number of clusters was selected based on the elbow method [45], the main idea of which is to define clusters such that the total intra-cluster variation (or total within-cluster sum of square (wss)) is minimized. As seen in Figure 3, the elbow of the curve is formed when the number of clusters is equal to 4.

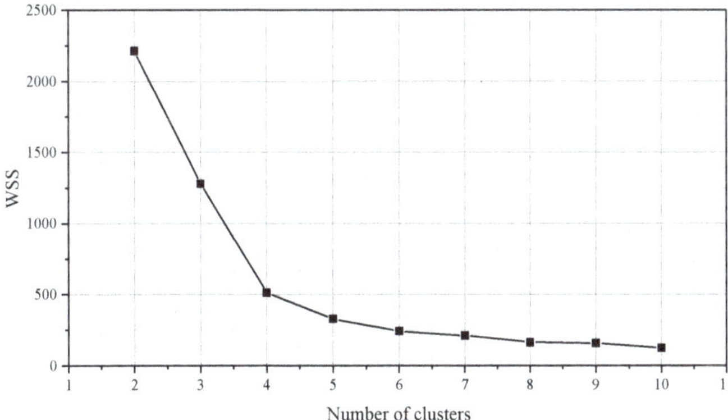

Figure 3. Within cluster sum of square dependence on the number of clusters.

3. Results

The most significant damage to the coast caused by hurricane "Anatol" was not a retreat of the shoreline, but a significant loss of sand from the protective dune ridge [48]. The cross-shore profile measured directly after the storm shows that significant volume of sand was lost from the underwater section of the beach, down to the effective closure depth of −5 m (Figure 4).

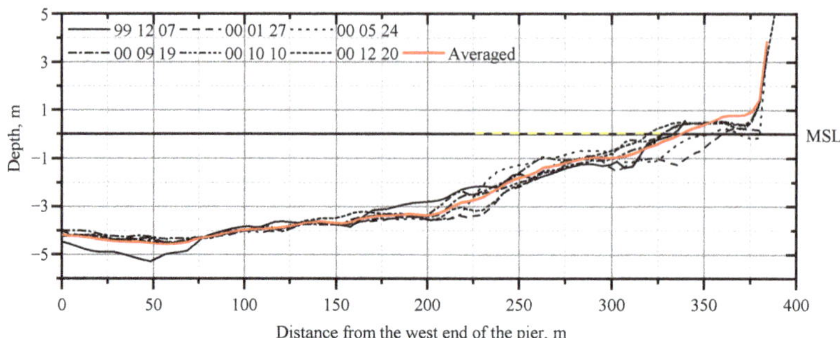

Figure 4. Changes in the beach profiles 1999–2000. The measurement date is indicated in the legend (yy mm dd); here and later, the yellow dash line indicates the length of the groyne.

Further erosion was observed in the month following the first survey on 7 December 1999, under relatively calm meteorological conditions with maximum 13 m/s predominant SW (22%) and W (23%) winds (Figure 5b). A total of 66 m^3 of sand per linear meter of shoreline was lost (Figure 6a(i)) in one month. The sand was moved from the −1.7 to −3.6 m depths to deeper than −4 m, covering the exposed moraine sediments. The position of MSL (0 m), moved landward by 20 m (Figure 6a(i)). The cross-shore sediment transport rate Q shows a bidirectional sediment transport tendency: shoreward at 0–180 m and seawards at 180–350 m resulting in accumulation of sand on the lower part of the profile (Figure 6b(i)).

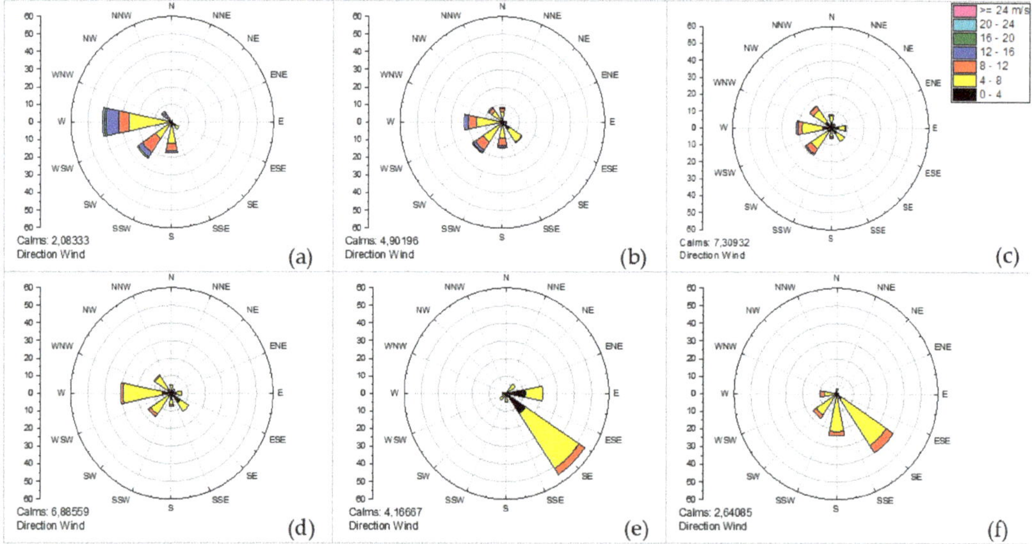

Figure 5. Wind roses for the study period: (**a**) 25 11 1999–06 12 1999; (**b**) 07 12 1999–26 01 2000; (**c**) 27 01 2000–23 05 2000; (**d**) 24 05 2000–19 09 2000; (**e**) 20 09 2000–09 10 2000; (**f**) 10 10 2000–19 12 2000.

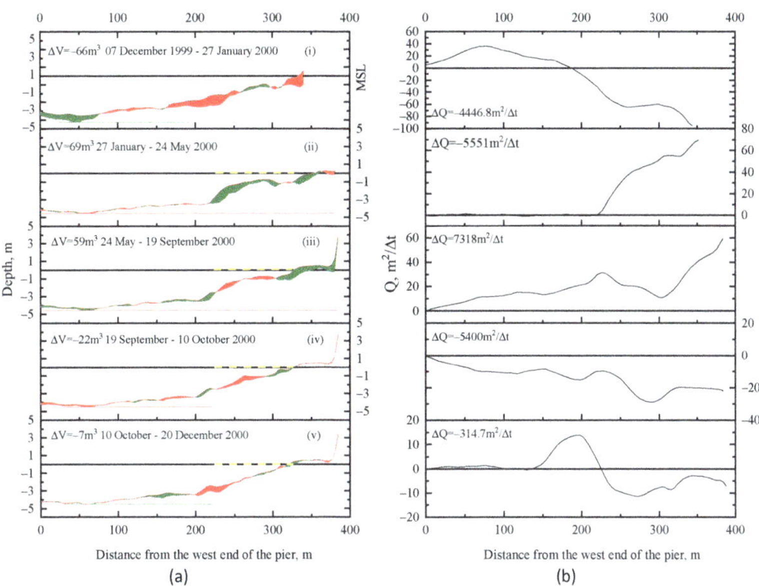

Figure 6. (**a**) Cross-shore profile changes (green-sand accumulation, red-sand loss) and (**b**) sediment transport rate Q, (**i**) December 1999–January2000, (**ii**) January–May2000, (**iii**) May–September 2000, (**iv**) September–October 2000, (**v**) October–December 2000.

From January to May 2000, there were several periods of strong wind with wind speeds up to 16 m/s, with westerly winds prevailing: 21% from W and 19% from SW directions (Figure 5c). These meteorological conditions led to coastal erosion: as a result, the groyne system (Figure 1b) was rebuilt [10]. Measurements in May 2000 showed that after the rebuilding of the groyne, 69 m^3 of sand per linear meter was accumulated. The largest accumulation took place on the upper part of the profile between 215 and 325 m (Figure 6a(ii)). The cross-shore sediment transport volume across the whole profile from January to May was 5551 m^2 with sand transport shoreward from the 218 m position (Figure 6b(ii)).

In summer (May–September 2020), calm weather prevailed (Figure 5d), and there was onshore transport. Almost the entire underwater profile shows positive Q transport, and, as a result, there was accumulation on the upper part of the profile, and the dune was partly replenished with sand (Figure 6a(iii)).The presence of the newly rebuilt groyne system, together with a calm autumn, extended a favorable condition for sand accumulation on the upper part of the cross-shore profile. September–December 2000 was calm with average wind speeds up to 12 m/s, with a predominant SE direction (Figure 5e,f). Profile changes showed sand movement from the shallow area to the offshore (Figure 6a(iv,v)), and in three months (October–December) only 7 m^3 per linear meter of sand was lost on the underwater profile, with sand simply being relocated on the profile. It is noticeable that the cross-shore profile May–December was relatively stable and probably approximated the equilibrium profile shape for this location.

Profile measurements were repeated after 16 years with the expectation that perceptible erosion on Palanga beach [12,31,33,34,49,50] would be reflected in the underwater cross-shore profile. In addition, we expected to observe short-term changes in the cross-shore profile development during the stormy autumn season. Cross-shore profile measurements were repeated once per week in October–December 2016 and December 2017. Both measurement seasons were similar, with average wind speeds not exceeding 12 m/s. The significant difference between the two study seasons was in predominant wind direction:

in 2016—N, E, and SE winds were prevalent, and in 2017—S, SW and W wind directions prevailed (Figure 7).

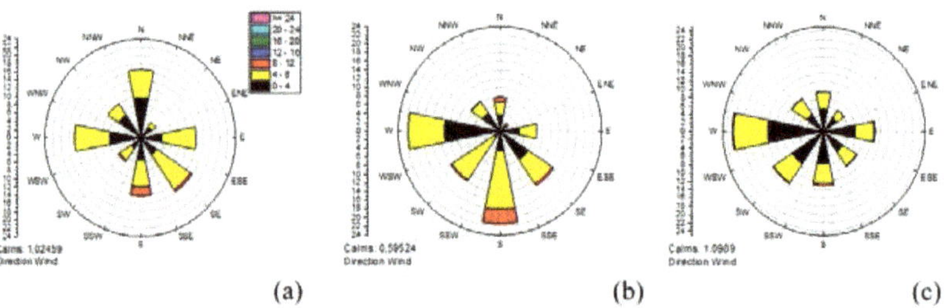

(a)   (b)   (c)

Figure 7. Wind roses for the October–December (**a**) 2016, (**b**) 2017 and (**c**) 2016 December–2017 October.

Over the sixteen years, 127 m³ per linear meter of sand accumulated on the cross-shore profile (Figure 8). Main changes were seen on the upper part of the profile, with the position of MSL moved seawards by 45 m. Net sediment transport rates were positive and indicated net sediment transport direction onshore. Dune expansion often creates the impression of beach narrowing and coastal erosion, which demonstrates the value of the data. Minor profile changes were seen in the deeper parts of the profile (perhaps indicating the real closure depth). Sediment was lost between 75 and 180 m with accumulation landward. The cross-shore sediment transport rate at a 75–300 m distance has negative values, showing a tendency for sand movement seawards. Sediment transport directed shoreward, Q, with positive values, is seen from 300 m. The sand accumulation zone starts from −4 m depth.

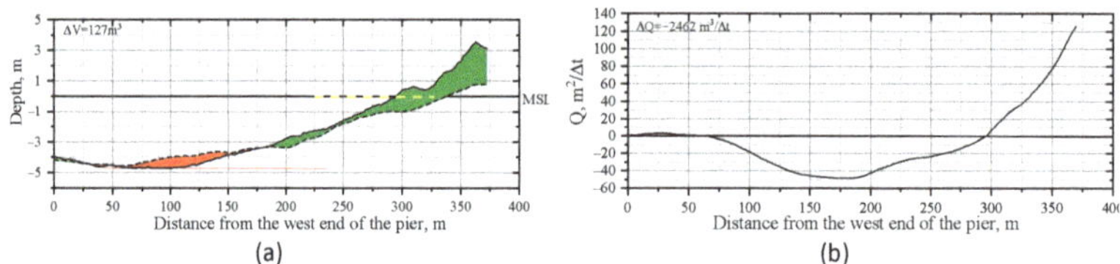

(a)   (b)

Figure 8. Cross-shore profile changes (green-sand accumulation, red-sand loss) (**a**) and sediment transport rate Q (**b**) over the period 2000 to 2016.

In 2016, changes in the cross-shore profile do not exceed 40 m³ per linear meter over one week. The sediment transport rate, Q, however, indicates considerable sand redistribution in the cross-shore profile without sand volume change (Figure 9). Noticeable changes in the profile shape were observed between 200 and 300 m and to −3 m depth. Small changes in weekly cross-shore profile volume show a quasi-stable state of the profile.

From December 2016 to October 2017, predominant wind directions were SW, NW, and W (Figure 7c). Wind speed did not exceed 12 m/s with the strongest winds from an E direction. Previous work [11] reported that westerly winds create favorable conditions for sand to remain on the underwater profile [11], but we observed sand loss under these conditions. In the relative calm and favorable wind direction conditions for accumulation, 36 m³ per linear meter of sand was eroded from location 200–300 m, between 0.5 m and −3 m depth (Figure 10a) and there was the retreat of the shoreline. In the dunes, some accumulation of sand occurred.

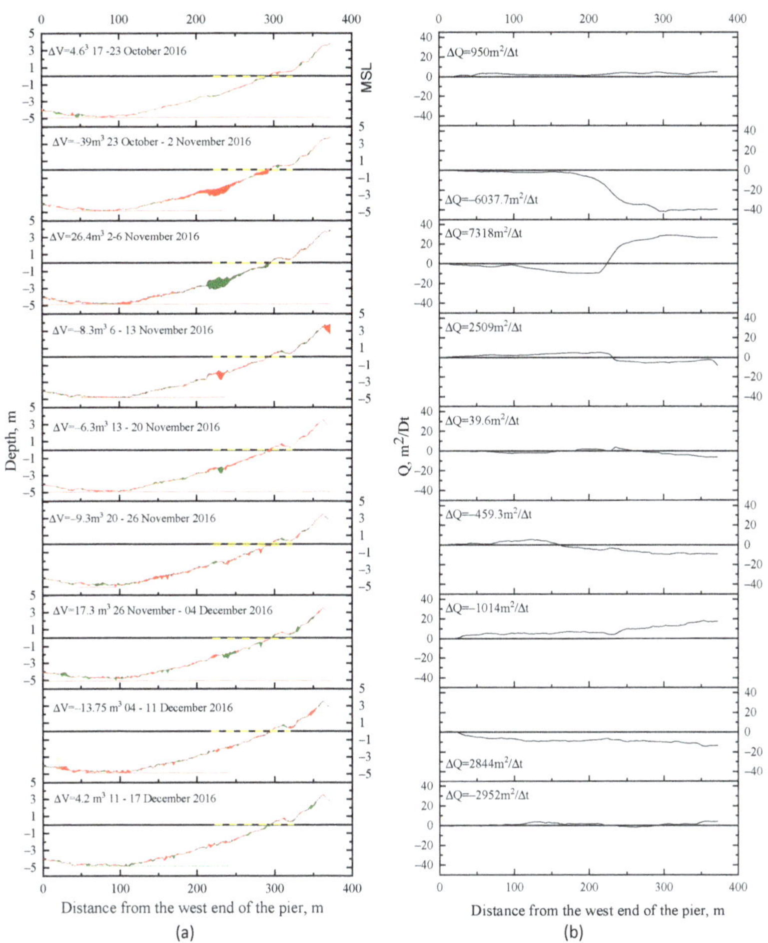

Figure 9. (**a**) Cross-shore profile changes (green-sand accumulation, red-sand loss) and (**b**) sediment transport rate Q in 2016.

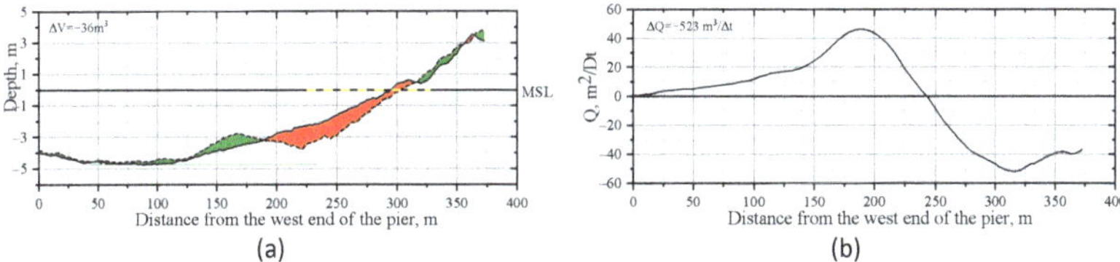

Figure 10. (**a**) Cross-shore profile changes (green-sand accumulation, red-sand loss) and (**b**) sediment transport rate Q, over the period 2016 to 2017.

A bi-directional cross-shore sediment transport rate Q structure was seen at that time (Figure 10b). Onshore sediment transport direction (Q positive) was observed from 0 m to 245 m and offshore sediment transport (Q-negative) from 245 to 375 m. As a result, a bar was formed at −3 m depth.

Short-term profile changes and cross-shore sediment transport tendencies in 2017 were similar to those observed in autumn 2016. The sand bar formed at −3 m remained stable during the observation period (Figure 11a). We captured very small sand volume relocations on the profile. Sand movement on the cross-shore profile was mostly observed between the bar and mean sea level position; the largest sand relocations were observed close to the MSL position.

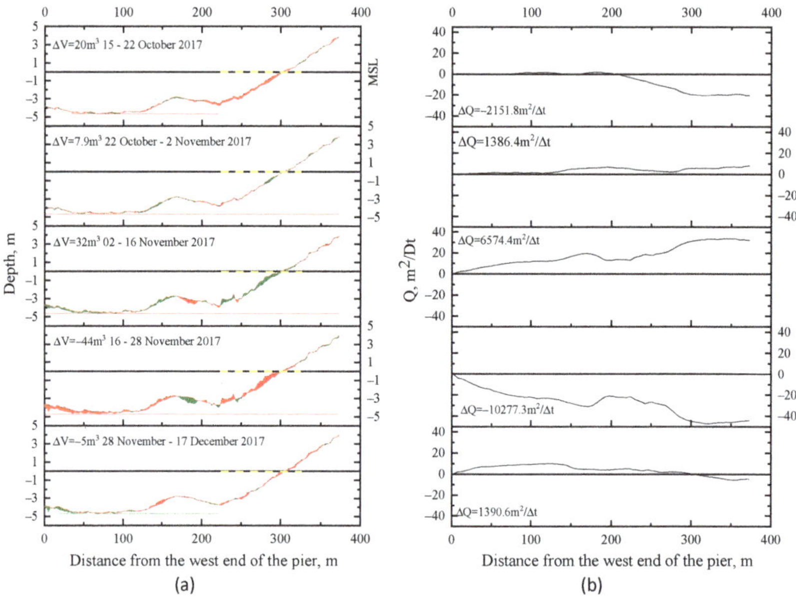

Figure 11. (**a**) Cross-shore profile changes (green-sand accumulation, red-sand loss) and (**b**) sediment transport rate Q, in 2017.

Total sand volume change during autumn 2017 was -30 m^3 per linear meter, and the cross-shore sediment transport rate was $\pm$ 4356.1 m^2 per week (Figure 11b). This tendency indicates sand relocation on the cross-shore profile without significant changes to the profile volume.

Seabed elevation changes in comparison with the average overall measured profiles indicate two different states (Figure 12). The seaward 130 m of the studied profile had minor changes in seabed elevation over 2016 and 2017. On more landward sectors, opposite trends were observed, being more active, with higher Δz values (up to $\pm$ 1 m). Only sectors greater than 300 m from the west end of the promenade bridge show positive Δz due to dune recovery after damage in December 1999.

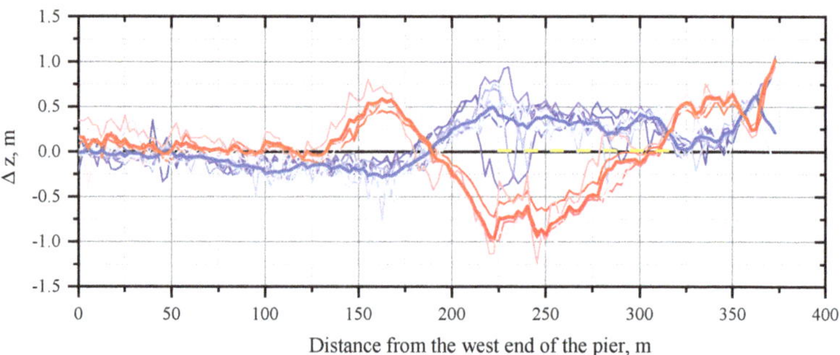

Figure 12. Spatial and temporal evolution of bottom elevation changes (Δz) showing the average of all measured profiles (blue—2016, red—2017).

To group cross-shore profile zones, K-means cluster analysis was performed. Cross-shore profile positions over the study period were grouped into four clusters, and all cross-shore profiles were also averaged. Standard deviations for the averaged curve were calculated (Figure 13).

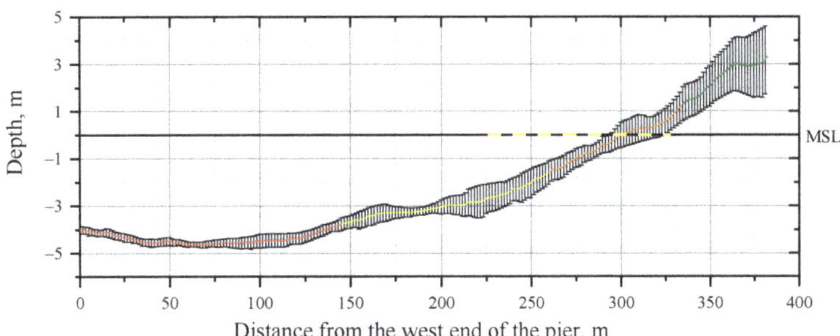

Figure 13. Average of all measured profiles (red 1st, yellow 2nd, orange 3rd and green 4th clusters) with the standard deviation.

The four clusters represent sectors on the cross-shore profile that change due to different conditions. The first cluster is the lower part of the profile from −5 m to 3.5 m and includes 38% of the total profile length (146 m). The deepest part of the profile is constrained by a moraine layer which is exposed in intense storms. Small scale seabed features result in minor (± 0.1 to ± 0.3 m) deviations from the average profile (Figure 14).

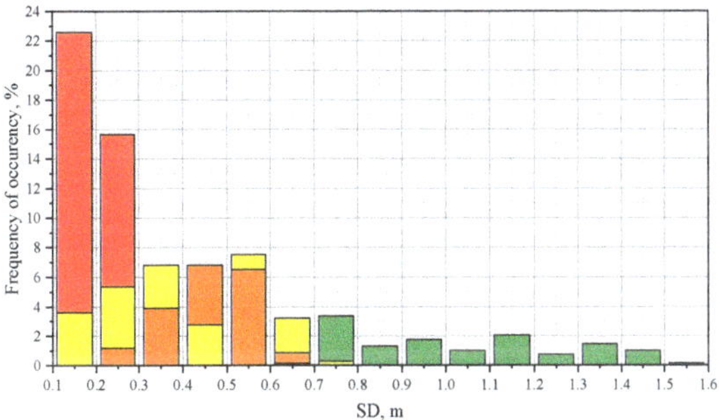

Figure 14. Distribution (%) of standard deviation (sd) per the clusters (red 1st, yellow 2nd, orange 3rd and green 4th clusters).

The second cluster comprises 114 m of the cross-shore profile, 30% of the length. It includes the middle section of the underwater profile from −3.5 m to −1.6 m depth. This part of the profile is above the depth of closure and falls into a more active hydrodynamic zone with more soft sediment than in deeper water. The standard deviation varies from ± 0.1 m to ± 0.8 m (Figure 14). The difference between the second and third clusters is likely the result of different hydrodynamic drivers. The second cluster is where waves and alongshore currents dominate most of the time, with the third cluster having similar process drivers, but affected by changes in the mean water level. This 73 m of the profile (19%) has a narrower distribution with standard deviation from ± 0.2 m to ± 0.7 m. The fourth cross-shore profile sector with the highest SD (from ± 0.2 m to ± 1.5 m), is the dune system, with different processes operating. It is the shortest cluster, just 49 m, with a height from 1.8 m to 3.2 m. The highest dune point was measured at 4.8 m in 2017.

4. Discussion

Beach profile features and evolution are essential considerations for coastal engineering projects. In the eastern Baltic Sea, profile dynamics have received comparatively less attention than alongshore processes, and the available knowledge remains partly qualitative and empirical [1,10,51–54].

Cross-shore profile evolution is often studied in a controlled environment such as in wave tanks [2,4–6], and results may be difficult to apply to realistic situations [9]. Often, the emphasis has been placed on the equilibrium profile for a particular location and the closure depth [51,55] but local geomorphological conditions (such as, in this case, the presence of a hard layer at −5 m) may be locally fundamental. In addition, interaction with structures on the coast may drastically change underwater beach profile evolution [6].

At Palanga beach, a quasi-equilibrium state [51] was reached in the hundred years after the promenade bridge was built [31,56]. This fragile state was destroyed in 1999 when a new pier, with an open design, was built [56], and substantial damage to the coast was done during storm "Anatol" [31,56]. Disruption of the quasi-equilibrium state created conditions that led to a new profile shape.

One storm, together with the new sediment transport conditions, changed the cross-shore profile and coastal zone characteristics enormously. Loss of sediment on the lower profile, deeper than −3 m, continued for another sixteen years. This area is further from the shoreline than the newly built groyne, but shallower than the depth of closure.

It is common practice to try to stop coastal erosion, using both soft (e.g., beach nourishment) and hard (e.g., wave breakers, groynes) coastal protection methods [57–60]. To stop further beach erosion, a groyne system was reconstructed in May 2000. First measurements after the groyne installation showed positive changes on the upper part of the profile. In sixteen years, 172 m^3 of sand per linear meter was accumulated, with full dune system recovery. A narrowing of the beach occurred along with a foredune recovery [25,61,62]. The foredune will likely be eroded by waves when water levels with around a ten-year return period is reached, but it has had enough time to recover its long-term equilibrium shape [63]. Foredune recovery is a slow process that may take years to decades [64].

Part of the cross-shore profile corresponding to the dune sector falls into a separate fourth cluster, with the largest values of change and standard deviation. This section of the profile shows different development trends, dependent on the sediment characteristics rather than on the hydrometeorological conditions, which is intuitively predictable, but not statistically proven [61,63,64].

For the first cluster, close to closure depth, significant sediment relocation was observed after the storm "Anatol". This area recovered fast, with insignificant (0.1–0.2 m) changes during most of the study period. Except during major storms there is little change on this deepest section of the profile due to the predominant short period waves in the SE Baltic Sea [61,63,64].

The rest of the profile was further divided into two parts, the second and third clusters. These sections of the profile behaved similarly, and the distribution of standard deviations fell within the same limits. We believe that the profile was separated into two sections due to predominant external forces (wave breaker, wave set-up, wave run-up and sea-level fluctuations) influencing the cross-shore sediment transport [51,65,66]. There is a need for additional studies to determine further which driving mechanisms dominate on which profile segment.

Comparison of cross-shore profiles in 2016 and 2017 show the importance of the direction of the predominant wind for profile evolution. Moreover, even a small change in the predominant wind direction from the south to the west caused opposite seabed elevation changes. This supports previous observations concerning the importance of the wind (and therefore wave) direction to Baltic Sea coastal evolution [12,15,26,67,68].

5. Conclusions

Three main conclusions regarding the cross-shore profile evolution at Palanga, eastern Baltic Sea coast, can be drawn from the results of this study. Firstly, the cross-shore profile is limited by a layer of hard moraine sediments which is exposed in intense storms. Secondly, intensive sediment relocation occurs at depths shallower than −4 m, where the wind (and therefore wave) direction and short-term sea-level fluctuations are critical to profile change. Thirdly, accurate beach profile data collection and analysis are essential because visual observations cannot be relied upon. It has been frequently stated that Palanga beach has been eroding in recent years, but apparent erosion, evidenced by a narrowing upper beach, is caused by dune advance.

Author Contributions: Methodology, R.Ž.; formal analysis, V.K.; investigation, L.K.-R., and V.K.; data curation, L.K.-R., and V.K.; writing—original draft preparation, L.K.-R.; writing review and editing, K.E.P.; visualization, V.K. All authors have read and agreed to the published version of the manuscript.

Funding: The APC was funded by the Baltic Research Programme (EEA Financial Mechanisms 2014-2021) project "Solutions to current and future problems on natural and constructed shorelines, eastern Baltic Sea" (EMP480). Parnell was also supported by the Estonian Ministry of Education and Research (Estonian Research Council, institutional support IUT33-3), and the European Regional Development Fund program Mobilitas Pluss MOBTT72, No. 2014-2020.4.01.16-0024.

Institutional Review Board Statement: Not applicable.

Informed Consent Statement: Not applicable.

Data Availability Statement: The data presented in this study are available on request from the corresponding author.

Acknowledgments: The authors would like to thank Emilija Griniūtė for assistance with data collection.

Conflicts of Interest: The authors declare no conflict of interest.

References

1. Aagaard, T.; Hughes, M.G. Equilibrium shoreface profiles: A sediment transport approach. *Mar. Geol.* **2017**, *390*, 321–330. [CrossRef]
2. Baldock, T.E.; Birrien, F.; Atkinson, A.; Shimamoto, T.; Wu, S.; Callaghan, D.P.; Nielsen, P. Morphological hysteresis in the evolution of beach profiles under sequences of wave climates—Part 1; observations. *Coast. Eng.* **2017**, *128*, 92–105. [CrossRef]
3. Sanchez-Arcilla, A.; Caceres, I. An analysis of nearshore profile and bar development under large scale erosive and accretive waves. *J. Hydraul. Res.* **2017**, *56*, 231–244. [CrossRef]
4. Baldock, T.E.; Alsina, J.A.; Caceres, I.; Vicinanza, D.; Contestabile, P.; Power, H.; Sanchez-Arcilla, A. Large-scale experiments on beach profile evolution and surf and swash zone sediment transport induced by long waves, wave groups and random waves. *Coast. Eng.* **2011**, *58*, 214–227. [CrossRef]
5. Baldock, T.E.; Manoonvoravong, P.; Pham, K.S. Sediment transport and beach morphodynamics induced by free long waves, bound long waves and wave groups. *Coast. Eng.* **2010**, *57*, 898–916. [CrossRef]
6. Beuzen, T.; Turner, I.L.; Blenkinsopp, C.E.; Atkinson, A.; Flocard, F.; Baldock, T.E. Physical model study of beach profile evolution by sea level rise in the presence of seawalls. *Coast. Eng.* **2018**, *136*, 172–182. [CrossRef]
7. Austin, M.; Masselink, G.; O'Hare, T.; Russell, P. Onshore sediment transport on a sandy beach under varied wave conditions: Flow velocity skewness, wave asymmetry or bed ventilation? *Mar. Geol.* **2009**, *259*, 86–101. [CrossRef]
8. Kobayashi, N.; Zhu, T.; Mallavarapu, S. Equilibrium beach profile with net cross-shore sand transport. *J. Waterw. Port Coast. Ocean Eng.* **2018**, *144*. [CrossRef]
9. Patterson, D.C.; Nielsen, P. Depth, bed slope and wave climate dependence of long term average sand transport across the lower shoreface. *Coast. Eng.* **2016**, *117*, 113–125. [CrossRef]
10. Roelvink, J.A.; Stive, M.F.J. Bar-generating cross-shore flow mechanisms on a beach. *J. Geophys. Res.* **1989**, *94*, 4785–4800. [CrossRef]
11. Žaromskis, R. Impact of different hydrometeorological condition on Palanga shore zone relief. *Geografija* **2005**, *41*, 17–24.
12. Bagdanavičiūtė, I.; Kelpšaitė, L.; Daunys, D. Assessment of shoreline changes along the Lithuanian Baltic Sea coast during the period 1947–2010. *Baltica* **2012**, *25*, 171–184. [CrossRef]
13. Soomere, T.; Viška, M. Simulated wave-driven sediment transport along the eastern coast of the Baltic Sea. *J. Mar. Syst.* **2014**, *129*, 96–105. [CrossRef]

14. Knaps, R. Sediment transport in the coastal area of the Eastern Baltic. In *Development of Marine Coasts within the Conditions of Fluctuation Movements of the Earth Crust*; Valgus: Tallinn, Estonia, 1966.

15. Viška, M. *Sediment Transport Patterns along the Eastern Coasts of the Baltic Sea*; Tallin University of Technology: Tallinn, Estonia, 2014.

16. Žaromskis, R.; Gulbinskas, S. Main patterns of coastal zone development of the Curonian Spit, Lithuania. *Baltica* **2010**, *23*, 146–156.

17. Lapinskis, J. Coastal sediment balance in the eastern part of the Gulf of Riga (2005–2016). *Baltica* **2017**, *30*, 87–95. [CrossRef]

18. Bagdanavičiūtė, I.; Kelpšaitė-Rimkienė, L.; Galinienė, J.; Soomere, T. Index based multi-criteria approach to coastal risk assesment. *J. Coast. Conserv.* **2018**. [CrossRef]

19. Eberhards, G.; Grine, I.; Lapinskis, J.; Purgalis, I.; Saltupe, B.; Toklere, A. Changes in Latvia's seacoast (1935–2007). *Baltica* **2009**, *22*, 12.

20. Bagdanavičiūtė, I.; Kelpšaitė, L.; Soomere, T. Multi-criteria evaluation approach to coastal vulnerability index development in micro-tidal low-lying areas. *Ocean Coast. Manag.* **2015**, *104*, 124–135. [CrossRef]

21. Eberhards, G.; Lapinskis, J. Processes on the Latvian coast of the Baltic Sea. In *Atlas*; University of Latvia: Riga, Latvia, 2008; p. 64.

22. Bagdanavičiutė, I.; Kelpšaitė, L.; Daunys, D. Long term shoreline changes of the Lithuanian Baltic Sea continental coast. In Proceedings of the 2012 IEEE/OES Baltic International Symposium (BALTIC), Klaipeda, Lithuania, 8–10 May 2012; pp. 1–6.

23. Łabuz, T.A.; Grunewald, R.; Bobykina, V.; Chubarenko, B.; Česnulevičius, A.; Bautrėnas, A.; Morkūnaitė, R.; Tonisson, H. Coastal dunes of the Baltic Sea shores: A review. *Quaest. Geogr.* **2018**, *37*, 47–71. [CrossRef]

24. Povilanskas, R.; Riepšas, E.; Armaitienė, A.; Dučinskas, K.; Taminskas, J. Shifting Dune Typesof the Curonian Spit and Factors of Their Development. *Baltic For.* **2011**, *17*, 215–226.

25. Urboniene, R.; Kelpšaite, L.; Borisenko, I. Vegetation impact on the dune stability and formation on the Lithuanian coast of the Baltic sea. *J. Environ. Eng. Landsc. Manag.* **2015**, *23*, 230–239. [CrossRef]

26. Kelpšaitė, L.; Dailidienė, I. Influence of wind wave climate change to the coastal processes in the eastern part of the Baltic Proper. *J. Coast. Res.* **2011**, *64*, 220–224.

27. Soomere, T.; Räämet, A. Spatial patterns of the wave climate in the Baltic Proper and the Gulf of Finland. *Oceanologia* **2011**, *53*, 335–371. [CrossRef]

28. Pindsoo, K.; Soomere, T.; Zujev, M. Decadal and long-term variations in the wave climate at the Latvian coast of the Baltic Proper. In Proceedings of the Ocean: Past, Present and Future—2012 IEEE/OES Baltic International Symposium, BALTIC 2012, Klaipėda, Lithuania, 8–11 May 2012.

29. Suursaar, Ü.; Kullas, T. Decadal variations in waveheights off Cape Kelba, Saaremaa Island, andtheir relationships withchanges in wind climate. *Oceanologia* **2009**, *51*, 39–61. [CrossRef]

30. Povilanskas, R.; Armaitienė, A. Seaside resort-hinterland Nexus: Palanga, Lithuania. *Ann. Tour. Res.* **2011**, *38*, 1156–1177. [CrossRef]

31. Dubra, V. Influence of hydrotechnical structures on the dynamics of sandy shores: The case of Palanga on the Baltic coast. *Baltica* **2006**, *19*, 3–9.

32. Jarmalavičius, D. Sea Coast Dynamics Next to the Palanga in Last Century. Available online: https://www.lrt.lt/naujienos/tavo-lrt/15/47235/d-jarmalavicius-juros-kranto-ties-palanga-kaita-per-paskutini-simtmeti-radijo-paskaita (accessed on 29 November 2019).

33. Jarmalavicius, D.; Šmatas, V.; Stankunavicius, G.; Pupienis, D.; Žilinskas, G. Factors controlling coastal erosion during storm events. *J. Coast. Res.* **2016**, *1*, 1112–1116. [CrossRef]

34. Jarmalavičius, D.; Satkunas, J.; Žilinskas, G.; Pupienis, D. Dynamics of beaches of the Lithuanian coast (the Baltic Sea) for the period 1993-2008 based on morphometric indicators. *Environ. Earth Sci.* **2012**, *65*, 1727–1736. [CrossRef]

35. Ulbrich, U.; Fink, A.H.; Klawa, M.; Pinto, J.G. Three extreme storms over Europe in December 1999. *Weather* **2001**, *S56*, 10. [CrossRef]

36. Mäll, M.; Suursaar, Ü.; Nakamura, R.; Shibayama, T. Modelling a storm surge under future climate scenarios: Case study of extratropical cyclone Gudrun (2005). *Nat. Hazards* **2017**, *89*, 1119–1144. [CrossRef]

37. Fink, A.H.; Brücher, T.; Ermert, V.; Krüger, A.; Pinto, J.G. The European storm Kyrill in January 2007: Synoptic evolution, meteorological impacts and some considerations with respect to climate change. *Nat. Hazards Earth Syst. Sci.* **2007**, *9*, 405–423. [CrossRef]

38. Orviku, K.; Suursaar, Ü.; Tonisson, H.; Kullas, T.; Rivis, R.; Kont, A. Coastal changes in Saaremaa Island, Estonia, caused by winter storms in 1999, 2001, 2005 and 2007. *J. Coast. Res.* **2007**, *II*, 1651–1655.

39. Žilinskas, G. Distinguishing priority sectors for the Lithuanian Baltic Sea coastal management. *Baltica* **2008**, *21*, 85–94.

40. Stankevičius, A. *Conditions of the Baltic Sea environment*; JTD, A., Ed.; AAA JTD: Vilnius, Lithuania, 2013.

41. Žilinskas, G.; Pupienis, D.; Jarmalavičius, D. Possibilities of Regeneration of Palanga Coastal Zone. *J. Environ. Eng. Landsc. Manag.* **2010**, *18*, 92–101. [CrossRef]

42. United States Army Corps of Engineers. *CEM: Coastal Engineering Manual*; U.S. Army Corps of Engineers: Washington, DC, USA, 2002.

43. Soomere, T.; Viška, M.; Eelsalu, M. Spatial variations of wave loads and closure depths along the coast of the eastern Baltic Sea. *Est. J. Eng.* **2013**, *19*, 93–109. [CrossRef]

44. Soomere, T.; Männikus, R.; Pindsoo, K.; Kudryavtseva, N.; Eelsalu, M. Modification of closure depths by synchronisation of severe seas and high water levels. *Geo-Mar. Lett.* **2017**, *37*, 35–46. [CrossRef]

45. Laccetti, G.; Lapegna, M.; Mele, V.; Romano, D.; Szustak, L. Performance enhancement of a dynamic K-means algorithm through a parallel adaptive strategy on multicore CPUs. *J. Parallel Distrib. Comput.* **2020**, *145*, 34–41. [CrossRef]
46. Steinley, D.; Brusco, M.J. Initializing K-means Batch Clustering: A Critical Evaluation of Several Techniques. *J. Classif.* **2007**, *24*, 99–121. [CrossRef]
47. Melnykov, V.; Michael, S. Clustering Large Datasets by Merging K-Means Solutions. *J. Classif.* **2019**. [CrossRef]
48. Žilinskas, G.; Jarmalavičius, D.; Kulvičienė, G. Assessment of the effects caused by the hurricane 'Anatoli' on the Lithuanian marine coast. *Geogr. Metraštis* **2000**, *33*, 191–206.
49. Jarmalavičius, D.; Žilinskas, G.; Pupienis, D.; Kriaučiuniene, J. Subaerial beach volume change on a decadal time scale: The Lithuanian Baltic Sea coast. *Z. Geomorphol.* **2017**, *61*, 149–158. [CrossRef]
50. Pupienis, D.; Jarmalavičius, D.; Žilinskas, G.; Fedorovič, J. Beach nourishment experiment in Palanga, Lithuania. *J. Coast. Res.* **2014**, *70*, 490–495. [CrossRef]
51. Dean, R.G.; Dalrymple, R.A. *Coastal Processes with Engineering Applications*; Cambridge University Press: Cambridge, UK, 2002.
52. Miller, J.K.; Dean, R.G. A simple new shoreline change model. *Coast. Eng.* **2004**, *51*, 531–556. [CrossRef]
53. Sousa, W.R.N.d.; Souto, M.V.S.; Matos, S.S.; Duarte, C.R.; Salgueiro, A.R.G.N.L.; Neto, C.A.d.S. Creation of a coastal evolution prognostic model using shoreline historical data and techniques of digital image processing in a GIS environment for generating future scenarios. *Int. J. Remote Sens.* **2018**, 1–15. [CrossRef]
54. Jara, M.S.; González, M.; Medina, R.; Jaramillo, C. Time-Varying Beach Memory Applied to Cross-Shore Shoreline Evolution Modelling. *J. Coast. Res.* **2018**, *345*, 1256–1269. [CrossRef]
55. Dean, R.G. Equilibrium beach profiles: Characteristics and applications. *J. Coast. Res.* **1991**, *7*, 53–84.
56. Žilinskas, G. Trends in dynamic processes along the Lithuanian Baltic coast. *Acta Zool. Litu.* **2005**, *15*, 204–207. [CrossRef]
57. Gittman, R.K.; Fodrie, F.J.; Popowich, A.M.; Keller, D.A.; Bruno, J.F.; Currin, C.A.; Peterson, C.H.; Piehler, M.F. Engineering away our natural defenses: An analysis of shoreline hardening in the US. *Front. Ecol. Environ.* **2015**, *13*, 301–307. [CrossRef]
58. Summers, A.; Fletcher, C.H.; Spirandelli, D.; McDonald, K.; Over, J.-S.; Anderson, T.; Barbee, M.; Romine, B.M. Failure to protect beaches under slowly rising sea level. *Clim. Chang.* **2018**, *151*, 427–443. [CrossRef]
59. Armstrong, S.B.; Lazarus, E.D. Masked Shoreline Erosion at Large Spatial Scales as a Collective Effect of Beach Nourishment. *Earth's Future* **2019**, *7*, 74–84. [CrossRef]
60. Romine, B.M.; Fletcher, C.H. Hardening on eroding coasts leads to beach narrowing and loss on Oahu, Hawaii. In *Pitfalls of Shoreline Stabilization: Selected Case Studies*; Cooper, J., Andrew, G., Pilkey, O., Eds.; Springer Science and Business Media: Dordrecht, The Netherlands, 2012.
61. De Almeida, L.R.; González, M.; Medina, R. Morphometric characterization of foredunes along the coast of northern Spain. *Geomorphology* **2019**, *338*, 68–78. [CrossRef]
62. Jarmalavičius, D.; Pupienis, D.; Žilinskas, G.; Janušaitė, R.; Karaliūnas, V. Beach-Foredune Sediment Budget Response to Sea Level Fluctuation. Curonian Spit, Lithuania. *Water* **2020**, *12*, 583. [CrossRef]
63. Pellón, E.; de Almeida, L.R.; González, M.; Medina, R. Relationship between foredune profile morphology and aeolian and marine dynamics: A conceptual model. *Geomorphology* **2020**, 351. [CrossRef]
64. Castelle, B.; Bujan, S.; Ferreira, S.; Dodet, G. Foredune morphological changes and beach recovery from the extreme 2013/2014 winter at a high-energy sandy coast. *Mar. Geol.* **2017**, *385*, 41–55. [CrossRef]
65. Komar, P.D. Coastal erosion processes and impacts: The consequences of Earth's changing climate and human modifications of the environment. *Earth Syst. Environ. Sci.* **2011**, 285–308. [CrossRef]
66. Komar, P.D. *Beach Processes and Sedimentation*; Prentice Hall: Upper Saddle River, NJ, USA, 1998.
67. Viška, M.; Soomere, T. Simulated and observed reversals of wave-driven alongshore sediment transport at the eastern baltic sea coast. *Baltica* **2013**, *26*, 145–156. [CrossRef]
68. Soomere, T.; Bishop, S.R.; Viška, M.; Räämet, A. An abrupt change in winds that may radically affect the coasts and deep sections of the Baltic Sea. *Clim. Res.* **2015**, *62*, 163–171. [CrossRef]

Journal of
Marine Science and Engineering

Gravel Barrier Beach Morphodynamic Response to Extreme Conditions

Kristian Ions [1], Harshinie Karunarathna [1,*], Dominic E. Reeve [1] and Douglas Pender [2]

[1] Zienkiewicz Centre for Computational Engineering, Faculty of Science and Engineering, Bay Campus, Swansea University, Fabian Way, Swansea SA1 8AA, UK; 920039@swansea.ac.uk (K.I.); d.e.reeve@swansea.ac.uk (D.E.R.)
[2] JBA Consulting, Unit 2.1, Quantum Court, Heriot Watt Research Park, Edinburgh EH14 4AP, UK; Doug.Pender@jbaconsulting.com
* Correspondence: h.u.karunarathna@swansea.ac.uk

Abstract: Gravel beaches and barriers form a valuable natural protection for many shorelines. The paper presents a numerical modelling study of gravel barrier beach response to storm wave conditions. The XBeach non-hydrostatic model was set up in 1D mode to investigate barrier volume change and overwash under a wide range of unimodal and bimodal storm conditions and barrier cross sections. The numerical model was validated against conditions at Hurst Castle Spit, UK. The validated model is used to simulate the response of a range of gravel barrier cross sections under a wide selection of statistically significant storm wave and water level scenarios thus simulating an ensemble of barrier volume change and overwash. This ensemble of results was used to develop a simple parametric model for estimating barrier volume change during a given storm and water level condition under unimodal storm conditions. Numerical simulations of barrier response to bimodal storm conditions, which are a common occurrence in many parts of the UK were also investigated. It was found that barrier volume change and overwash from bimodal storms will be higher than that from unimodal storms if the swell percentage in the bimodal spectrum is greater than 40%. The model is demonstrated as providing a useful tool for estimating barrier volume change, a commonly used measure used in gravel barrier beach management.

Keywords: gravel barrier beaches; storms; XBeach; morphodynamic change; overwash; bimodal spectrum

Citation: Ions, K.; Karunarathna, H.; Reeve, D.E.; Pender, D. Gravel Barrier Beach Morphodynamic Response to Extreme Conditions. *J. Mar. Sci. Eng.* **2021**, *9*, 135. https://doi.org/10.3390/jmse9020135

Academic Editor: Yoshimichi Yamamoto
Received: 17 December 2020
Accepted: 20 January 2021
Published: 28 January 2021

Publisher's Note: MDPI stays neutral with regard to jurisdictional claims in published maps and institutional affiliations.

1. Introduction

Gravel beaches and barriers form a significant proportion of world's beaches at mid to high latitudes [1]. They act as natural means of coast protection and are capable of dissipating a large portion of incident wave energy under highly energetic wave conditions (e.g., [2,3]). Gravel beach and barrier morphodynamics is dominated by the highly reflective nature of steep beach face, energetic swash motions generated by waves breaking on the lower shoreface and potential overwash of the beach crest [4]. Overwashing and overtopping of gravel beaches and barriers can occur during extreme storm conditions and can lead to crest build-up, crest lowering, landward retreat and potentially breaching [1,5,6].

Studies of gravel beach and barrier morphodynamics date back to a few decades. Powell [7] investigated the hydraulic behaviour of gravel beaches using a set of physical model testing results. Following that, Powell [8] presented a parametric model for gravel beach morphodynamic evolution against short term wave attack, based on a comprehensive series of physical model tests. The application of his model to a number of field sites proved it to be a useful tool to determine short term cross-shore profile evolution of gravel beaches. However, Powell's [8] model does not consider barrier overwash thus limiting its application to ordinary wave conditions. Bradbury [9] investigated a relationship between incident wave conditions and the geometry of barrier beaches under storm wave conditions using an extensive series of physical model experimental results on cross-shore profile change of gravel beaches. He developed an empirical dimensionless threshold called 'barrier inertia parameter', which is a function of emergent barrier cross-sectional area, freeboard and the incident wave height, to detect barrier breaching. Bradbury et al. [6] applied the Bradbury [9] empirical parameter to detect the morphodynamic response of a number of barrier beaches in the southern England. The barrier inertia parameter is found to be a useful tool to identify incident wave conditions leading to barrier beach although the parameter is not able to predict morphodynamic change of the barrier. Also, the range of validity of it is limited to incident wave steepness below a certain value and extrapolation of the parameter to higher incident wave steepness has found to be problematic and unreliable [6].

Most investigations on gravel beach and barrier overwash and morphodynamics found in literature are based on either experimental investigations e.g., [7–13], or field studies e.g., [1,3,4,14–16]. Some attempts have also been made to apply numerical models to simulate morphodynamic response to incident waves and water levels. Williams et al. [17] used XBeach process-based coastal morphodynamic model [18] to investigate overwashing and breaching of a cross-shore profile of a gravel barrier located in a macrotidal environment in the south-west coast of the UK. They had some success but, the model, which was developed for sandy beaches, over-predicted the erosion of the upper beach face although the model was able to identify the threshold water level and wave conditions for overwashing. Jamal et al. [19] modified the XBeach model to investigate the morphodynamic behaviour of gravel beaches by introducing a coarse sediment transport formula and groundwater infiltration/exfiltration phenomena. The modified model (XBeach v12) was found to capture gravel transport and beach morphodynamics satisfactorily. McCall et al. [20] also presented an extension to XBeach to simulate morphodynamics of gravel beaches. This model, called XBeach-G, captured berm building and roll-over of gravel barriers. Gharagozlou et al. [21] used the XBeach-G model to simulate overwash, erosion and breach of barrier islands during extreme storm conditions. Subsequently, Phillips et al. [22] used the non-hydrostatic version of XBeachX [23] to study intertidal foreshore evolution and runup of gravel barriers. They concluded that the sandy intertidal area in their study site plays an important role in runup and overwash of gravel barriers.

Critical to the need to investigate barrier beach morphodynamics is the ultimate objective for practitioners to make sustainable long-term decisions on the management of these systems. As the climate continues to change, multiple questions arise that require consideration for coastal managers. E.g., when should failure/breaching of barrier beach be managed? Can this system function effectively as flood/erosion protection in the future? How might the barrier have to be adapted to continue to manage risk? Are long-term adaptation requirements technically and economically feasible? In the UK, the above type of decision making is typically made assuming a 100-year period for new interventions. Given the inherent uncertainty in all aspects of this process it is important that coastal engineers have tools available that can quickly support decision making on the critical aspects and provide focus for more detailed studies if required.

This paper aimed at answering some questions laid out above and addressing two research needs identified in literature: (i) Although process-based models are useful tools to investigate gravel beach and barrier response to extreme conditions, establishing a high-resolution numerical model to a given site can be time-consuming and costly. Also, significant uncertainties surround deterministic simulations due to uncertain input parameters and inadequate process-descriptions in those models thus requiring a large number of simulations to quantify uncertainties. Therefore, the application of these models may have limited scope within the coastal engineering industry where time and resources are restricted and expensive; (ii) Simple gravel barrier models have their own limitations: Powell's [8] parametric model is not able to capture barrier overwash. The barrier inertia parameter defined by Bradbury et al. [6] links barrier inertia into breaching thresholds but does not capture morphodynamic change. Poate et al. [24] was successful with an attempt to parameterise wave runup on a gravel beach but did not investigate gravel barrier morphodynamics.

The first part of this paper will address the need to develop a simple parametric model to capture gravel barrier beach morphodynamic response to extreme conditions. The second part is devoted to gravel barrier response to bimodal storm conditions. Although gravel beach response to waves has been studied extensively, for regular and irregular waves, there is little reported investigation of their response to bimodal wave conditions, quite commonly experienced at midlatitudes. Bradbury et al. [25] highlight the damaging effects of bimodal storms on gravel barriers, which was consolidated by Thompson et al. [26] and a few others. The results of barrier response to bimodal storms will be compared with that of unimodal storms.

A large ensemble of numerically simulated barrier evolution and overwash results, taking the Hurst Castle Spit gravel barrier beach located in the south coast of the UK as the study site, is used to develop the parametric model in the first part of the paper. The same study site is used in the second part of the paper to investigate the impacts of bimodal storms on gravel barrier morphodynamics.

2. Hurst Castle Spit

Hurst Castle Spit (HCS) gravel barrier beach system forms the Christchurch Bay and provides protection from wave attack to an extensive area of low-lying land in the Western Solent in the south coast of the UK [5,27] (Figures 1 and 2).

Figure 1. Location of Study, Hurst Spit Beach SW England located in the English Channel, UK (Google Earth).

The Spit is approximately 2.5 km long, orientated 130° North. The beach foreshore has an average slope of 8°, with crest height varying significantly along the beach, ranging from 7 m–3 m mODN, from East to West, where ODN is 'Ordnance Datum Newlyn' which refers to height of mean sea-level in the United Kingdom, with reference to a location named Newlyn [28].

The Spit is mostly shingle composition, with sediment diameter varying between 6 mm and 45 mm, with the mean sediment diameter D_{50} of 15 mm and D_{90} of 45 mm [5,29]. It is estimated that the main body of the HCS is declining in volume by approximately 7000–8000 m^3/yr and retreating by 3.5 m/yr on average [27]. The littoral transport rate in the nearshore region has been estimated at 11,000–13,000 m^3 per/yr [30]. Littoral transport results in accumulation of sediment on the eastern tip of the spit, towards the Hurst Castle (Figure 2) [30].

Figure 2. (**Top image**)—Hurst Spit aerial photograph, showing salt marshes and Hurst Castle located at the end of the spit (Channel Coastal Observatory). (**Bottom image**)—Hurst Spit Beach, showing gravel composition of the beach.

The tidal range around the HCP is 2.2 m where the spit is subjected to a meso-tidal regime. The predominant wave incidence is from the SSW. The Offshore bathymetry is complex. Shingles Bank is located offshore of the HCP, which is exposed at low tide [31]. The North Heads bank runs parallel to the shoreline. The banks have a sheltering effect on the spit from on-coming wave attack (Figure 3).

HCS is part of three international nature conservation designations, meaning it is of considerable environmental and geological interest in developing an understanding of coastal geomorphology. Numerous overwash and breaching events of HCS have been reported over the past 200 years, prior to the implementation of a management plan in 1996–1997. The spit underwent a major recharge in 1996 as part of a 50-year shoreline management plan. Despite having an artificial beach crest following beach recharge, the spit is still prone to severe damage from storms, most recently seen in February 2020 and in much greater effect during the 2013/2014 winter storms.

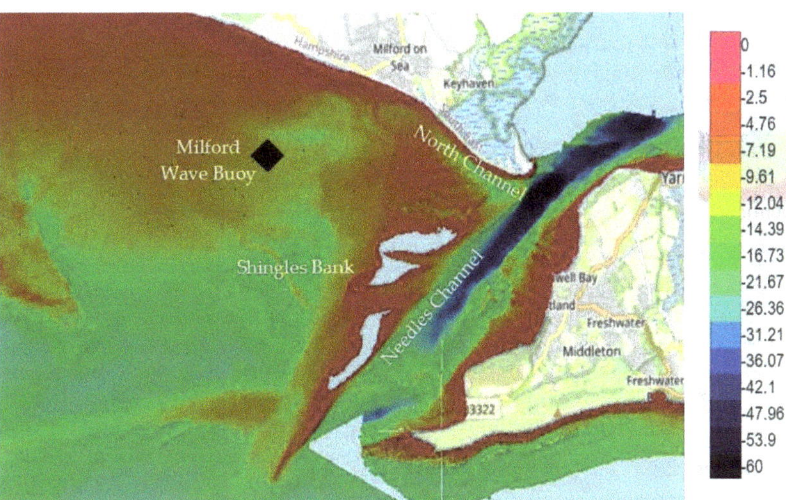

Figure 3. Nearshore bathymetry of Hurst Castle Spit gravel barrier beach and the surroundings. Bathymetry data are from Channel Coastal Observatory.

A wave buoy is situated to the East of the HCS, at a water depth of 10 m–12 m ODN [9] (Figure 3). The waves have been measured since 1996 at 1 Hz from which significant wave height, peak wave period and mean wave period have been calculated every 30 min [32]. Using the historical wave buoy data from the buoy the annual average significant wave height (H_s), average peak wave period (T_p) and average wave direction have been determined to be 0.65 m, 8.2 s and 211° respectively. Through nearshore wave modelling and studying previous literature Bradbury and Kidd [29] found that the maximum significant wave height varies between 3.57 m (240°) and 2.89 m (210°) annually on the eastern end of the spit and between 2.10 m (210°) and 2.68 m (240°) at the western end of the spit. Bradbury and Kidd [29] suggest that the mean value of the maximum nearshore wave height declines along spit from the east to the west due to the attenuating or dissipating influence of the Shingles Bank and the North Head Bank, resulting complex wave refraction and wave train "crossover". Wave shoaling and breaking (at low water) induced by the complex bathymetry of the banks and channels seawards of the spit reduces the height of offshore waves by almost one third [29]. The spit is highly vulnerable to high energy waves travelling across the Atlantic Ocean [27].

The south-west of the UK where the HCS is located is also subjected to frequent storms with bimodal characteristics as a result of swell-dominated waves with peak wave periods exceeding 16 s reaching from the Atlantic [26]. Bradbury et al. [25] found that bimodal conditions occur 25% of time during winter months where storms are frequent and severe. Nicholls and Webber [27] and Thompson et al. [26] and some others reveal that bimodal storms may induce greater beach erosion and damage during storms than their unimodal counterparts at certain occasions in the south and south-west of the UK. The swell percentage in the bimodal wave spectrum had found to exceed 50% at certain instances. When these swell conditions combine with mean high-water springs and storm surges, the chance of overwash of barrier beaches is found to be significantly increased [33].

3. Model Development and Numerical Simulations

To develop a reliable parametric model of gravel barrier beach change needs a substantial amount of data. As the availability of field or laboratory data covering a wide range of beaches and wave conditions are limited, we take the advantage of the open source process-based XBeach non-hydrostatic beach morphodynamic model (XBeachX) to generate a large number of realisations of overwashing and morphodynamic change of gravel barriers from a wide range of unimodal and bimodal storm conditions. XBeach model will be set up in 1D mode to a wide selection of cross sections of HCS to simulate overwash and barrier volume change.

3.1. XBeach Model

The open source, process-based XBeach model was developed by Roelvink et al. [18] as a sandy beach-dune erosion model. XBeach was originally developed as a short wave-averaged wave group resolving model and allowed short-wave variations in the wave group scale. The model resolved depth-averaged non-linear shallow water equation and the short wave motion is solved by the wave action balance equation using the HISWA equation [34].

Later, the XBeach model was extended to include a non-hydrostatic pressure correction [35,36] to the depth-averaged non-linear shallow water equation which allows modelling of the instantaneous water surface elevation. The depth-averaged dynamic pressure is derived using a method similar to a one-layer version of the SWASH model [37].

The development of XBeach [18,35] and XBeach-G [19,20,38,39] led to wide ranging studies on gravel beaches and barriers. Williams et al. [17] used early development of XBeach to compare the results to experimental data on gravel profile development and found the model accurately reproduced the results of erosion with accuracy Briers Skill Score (BSS) of 0.6 [40], however the model underpredicted the location and size of the storm berm. McCall et al. [20,39] and Masselink et al. [41] used XBeach-G to predict the morphological response of gravel beaches and compared them with Bradbury et al., [6] barrier inertia model. Masselink et al. [41] highlighted the importance of Bimodal wave spectrum on increased overwash events, which is further supported by Orimoloye et al. [42] and Thompson et al. [26].

Williams et al. [33]) used XBeach to develop an empirical framework for modelling the response of high energy coastlines. They demonstrated that XBeach can effectively model the breaching and overwash of barrier island with 70% of BSS scores ranging from good to excellent (0.6–0.8). Another comprehensive study using XBeach to derive an empirical expression for the overwash and runup on gravel beach has been presented by Poate et al. [24]. Their results were capable of improving run-up estimations on gravel beaches. Pairing of XBeach model-derived synthetic data with field data has been used by Almeida et al. [43], to find that spectral shape of wave forcing conditions plays a key role in morphological response of gravel barriers. Several other papers have used XBeach to model, with large success the processes of gravel beaches and barriers.

The non-hydrostatic XBeach model (named XBeachX) includes provisions for applications to gravel beaches [20,39]. It also has a ground water model that allows infiltration-exfiltration through the permeable gravel bed, which is a key process contributing to gravel beach morphodynamics and, gravel transport formulation of van Rijn [44] and numerous other sediment transport formulations [20,43,45].

The hydrodynamic equations solved by the 1D XBeach non-hydrostatic model are:

$$\frac{\partial \eta}{\partial t} + \frac{\partial h u}{\partial x} = 0 \tag{1}$$

$$\frac{\partial u}{\partial t} + u\frac{\partial u}{\partial x} - \frac{\partial}{\partial x}\left(v_h \frac{\partial u}{\partial x}\right) = -\frac{1}{\rho}\frac{\partial(\rho p_{nh} + \rho g \eta)}{\partial x} - \frac{\tau_b}{h} \tag{2}$$

where t is time, η is the water surface elevation from the still water level, u is the depth averaged cross-shore velocity, h is the total water depth, v_h is the horizontal viscosity, ρ is the density of seawater, p_{nh} is the depth averaged dynamic pressure normalised by the density, g is the gravitational acceleration and τ_b is the total bed shear stress given by:

$$\tau_b = \rho c_f u|u| \tag{3}$$

in which c_f is the dimensionless friction coefficient.

Transport of coarse sediment is calculated using Shields parameter:

$$\theta = \frac{\tau_b}{\rho g \Delta_i D_{50}} \tag{4}$$

where τ_b is bed shear stress and D_{50} is mean sediment diameter and Δ_i is the relative effective weight of the sediment. Equation (4) is used to calculate sediment transport using the bed load transport equation of van Rijn [44], excluding silt sediment:

$$q_b = \gamma D_{50} D^{-0.3} \sqrt{\frac{\tau_b}{\rho}} \frac{(\theta' - \theta_{cr})}{\theta_{cr}} \frac{\tau_b}{|\tau_b|} \tag{5}$$

q_b is the volumetric bed load transport rate, γ is the Van Rijn calibration coefficient set at 0.5 [44] and θ' is critical shields parameter. D is non-dimensionalized grain size and θ_{cr} is critical shields parameter, initiating sediment transport.

Bed level change is then calculated in the centre of each grid cell, using the Exner equation:

$$\frac{\partial \zeta}{\partial t} + \frac{1}{(1-n)} \frac{\partial q_b}{\partial x} = 0 \tag{6}$$

where ζ is the elevation of the bed above a horizontal datum and n is the porosity.

If the gradient of the beach $\lceil tan\ \beta \rceil$ is greater than the angle of repose θ, avalanching occurs and alters the beach profile shape.

The reader is referred to McCall et al. [20,39] and XBeach Manual [46] for full details of the model.

3.2. XBeachX Model Calibration and Validation

Prior to the application of the XBeachX model for gravel beaches to generate a synthetic series of gravel barrier overwash and profile evolution from extreme conditions, the model was calibrated using field measurements at the HCS. The primary focus of this study was morphological change of the barrier crest above 0 mODN. A numerical XBeachX model was first established to HCS in 1D form, to two cross sections (Figure 4) of the barrier beach using measured pre-storm barrier cross-sections. Table 1 shows cross-sections and storm conditions used for model calibration. The selected cross-sections vary in size, shape and crest height and have different degrees of susceptibility to storm erosion. The model domain was extended until the 15 m water depth using a 1:50 beach slope [46], which is sufficiently steep to have no effect on waves at the model boundary in the subtidal region to ensure no wave transformation between Milford wave buoy and the model boundary. The offshore and nearshore bathymetry required for numerical model domain development were obtained from the bathymetry measurements of the Channel Coastal Observatory (CCO) [32] of the UK. A 1D, non-equidistant grid system, oriented with the x-axis in the cross-shore direction along the cross-section, positive towards the shoreline, was used. The grid cell size varied from 2 m–3 m offshore to 0.1 m–0.3 m onshore, which allowed the model to capture the complex morphodynamics of the beach cross section, whilst offshore grid sizes are sufficient in size to capture wave transformations occurring in the non-hydro static model.

The input storm wave boundary conditions were derived from the wave data from the Milford wave buoy (50°42′75 N, 01°36′91 W) located 10 m water depth off the coast of Milford on Sea. The wave buoy is located to the left side of the shingle bank and the North Head Bank (Figure 3). The two selected storms vary considerably in their characteristics where S1 dominated by long distance swell approaching from the Atlantic Ocean and show evidence of strong bimodal seas while S2 is dominated by local wind waves. It should be noted that the selection of storm conditions used for model calibration was limited by the availability of accurate pre- and post-storm beach profile measurements at HCS. Water levels during the storms were derived from the nearest tide gauge located in Christchurch Bay, from the UK tide Gauge Network of the British Oceanographic Data

Centre (BODC) [28]. Storms were determined based on the storm threshold wave height defined by the CCO for south-west region of the UK where waves are considered as storm waves if the significant wave height exceeds the 0.25-year return period significant wave height [32]. Storm surge was derived from Coastal flood boundary conditions for UK mainland and islands (DEFRA 2018) [47].

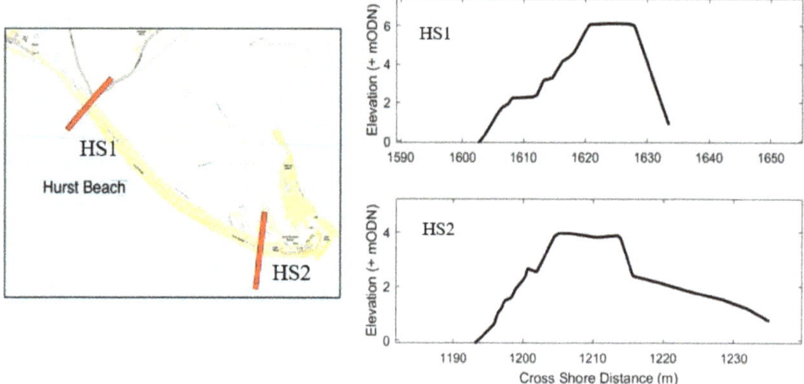

Figure 4. Cross sections of the Hurst Castle Spit gravel barrier beach used for model calibration and their locations. Locations (**left**) and cross section profiles (**right**), where left side of the figures are seaward and the right side of the figure is landward.

Table 1. HCS gravel barrier cross-sections and storm conditions extracted from the Milford wave buoy for model calibration, used for XBeach model validation. $(H_s)_{max}$ = maximum significant wave height; T_p = peak wave period from the JONSWAP spectrum; θ = wave approach direction from the north; MHWS = mean high water spring during the storm. S1—storm 1, S2—storm 2.

Cross Section	Crest Height m ODN	Pre-Storm Profile Date	Post-Storm Profile Date	Storm Duration (h)	$(H_s)_{max}$ (m)	T_p (s)	θ	MHWS above ODN (m)
HS1-S1	6.27	28/10/2011	31/10/2011	10 (S1)	2.75	18	220	1.1
HS1-S2	6.27	09/11/2011	13/11/2011	24 (S2)	3.85	8.3	216	0.928
HS2-S2	3.96	09/11/2011	13/11/2011	24 (S2)	3.85	8.3	216	0.928

Figure 5 shows a comparison of measured and XBeach-simulated post-storm cross-shore profile response of the HCS for storm conditions given in Table 1, following model calibration. The majority of the profile change in those cases are seen in the middle and lower part of the profile while the upper beach remains unchanged. This may be due to wave run-up not reaching the upper beach. The model used gravel beach sediment transport formulation given in McCall et al. [20] and D_{50} of 15 mm and D_{90} of 45 mm [29]. To accurately quantify the comparisons, the Briers Skill Score (BSS) [40] and RMSE are utilised. The BSS categorises the model's ability to correctly predict profile changes, where a score of 0–0.3 indicates 'poor' prediction, 0.3–0.6 indicates a 'reasonable/fair' model prediction, 0.6–0.8 indicates a 'good' score and lastly a score of 0.8–1.0 an excellent prediction. Both the skill score and RMSE were calculated using the profile change above 0 mODN due to the lack of accurate pre and post storm bathymetric measurements below 0 mODN available for storm calibration. Whilst morphological changes below the MSL dictate the evolution of the submerged beach step [15], the primary focus of this study was morphological change of the barrier crest above 0 mODN and therefore step dynamics are outside the scope of this study. The primary calibration parameters and the final selected values are given in Table 2. All selected parameter values were within the ranges recommended in the XBeach manual.

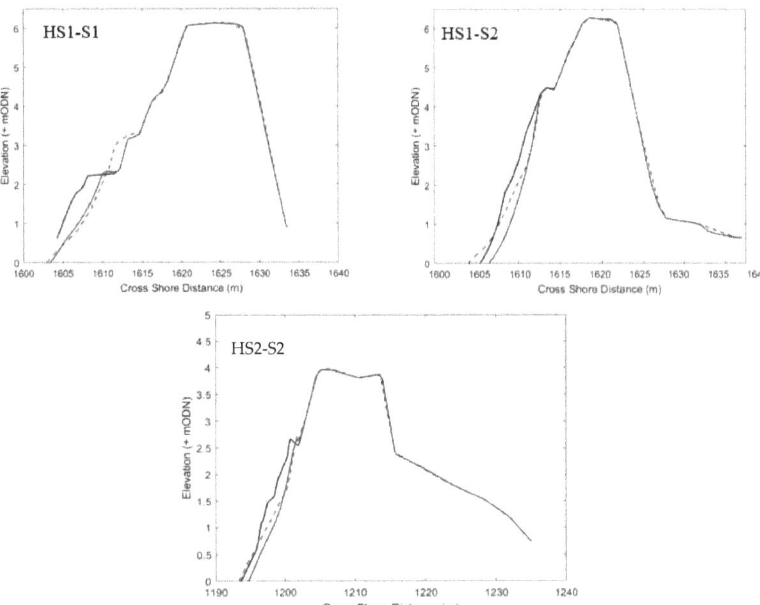

Figure 5. A comparison of measured and simulated post-storm profiles (HS1-top and HS2-bottom) at HCS following XBeachX model calibration against storms S1 and S2. Measured pre-storm profile (—black line), measured post-storm profile (—red dotted line) and simulated post-storm profile (—blue line).

Table 2. Calibration parameters of the XBeach non-hydrostatic model of the Hurst Castle Spit gavel barrier beach and the final selected values which gave the best fit profile with measured profile. dryslp is critical avalanching slope above water level, wetslp is critical avalanching slope below water level, CFL is maximum courant Friedrichs-Lewy number, repose angle is the angle of internal friction of sediment, kx is hydraulic gradient, ci is mass coefficient in shields inertia term, morfac is the morphological acceleration factor and cf is the bed friction factor.

Model Parameter	Recommended Range	Default Value	Selected Value
dryslp	0.1–2.0	1.0	1.0
wetslp	0.1–1.0	0.3	0.3
CFL	0.7–0.9	0.7	0.9
reposeangle	0–45	30	45
kx	0.01–0.3	0.01	0.15
ci	0.5–1.5	1.0	1.0
morfac	1–1000	1	1
Cf	3D90	3D90	3D90

The calibration results reveal that the model is capable of satisfactorily capturing the morphodynamic response of the HCS gravel barrier to storms. Profile change at HS1-S1 case scored BSS of 0.62 and RMSE 0.24. HS1-S2 scored BSS of 0.6 and RMSE of 0.14 while HS2-S2 scored BSS of 0.78 and RMSE of 0.094, providing either 'good' or 'excellent' predictions. It should be noted that the model underpredicted the berm formation in HS1-S1 (Figure 5.) but the erosion of the lower beach is captured well. The model slightly overpredicted the erosion of the intertidal zone in HS1-S2 but upper beach, which is the most important areas in terms of barrier area change, is correctly modelled. In HS2-S2, the accretion of sediment on the upper beach area is captured very well although there was some overprediction of profile erosion in the lower beach area.

Direct validation of the HCS model against barrier overwash was not possible due to lack of overwash data. Therefore, the model was used to simulate if overwash occur at HS1 and HS2, during a series of storms given in Table 3 [6].

Table 3. Storm conditions used to model barrier overwash for comparison with Bradbury et al. [6] results.

Storm Return Period	H_s (m)	T_m (s)	Storm Surge Imposed on MHWS (m)
1:1	3.69	8.64	1.0
1:10	4.22	9.30	1.0
1:20	4.39	9.48	1.0
1:50	4.6	9.71	1.0
1:100	4.75	9.87	1.0

Using the Barrier Inertia Model (BIM) developed by Bradbury [9] and Bradbury et al. [6] given in Equation (7), which states that if the barrier inertia parameter $\frac{R_c B_a}{H_s^3}$ is smaller than $0.0006 \left(\frac{H_s}{L_m} \right)^{-2.5375}$, overwashing will occur:

$$\frac{R_c B_a}{H_s^3} = 0.0006 \left(\frac{H_s}{L_m} \right)^{-2.5375} \quad \text{for } 0.15 < \frac{H_s}{L_m} < 0.032 \tag{7}$$

where R_c is barrier freeboard, B_a is barrier cross sectional area above MHWS + storm surge elevation (ESL), H_s is the significant storm wave height and L_m is deep water wavelength corresponding to mean wave period. Please refer to Figure 6 for definitions of variables. Using Equation (7), Bradbury et al. [6] estimated HS1 will not overwash during any of the storms given in Table 3 while HS2 will overwash during all storms except 1 in 1 year storm.

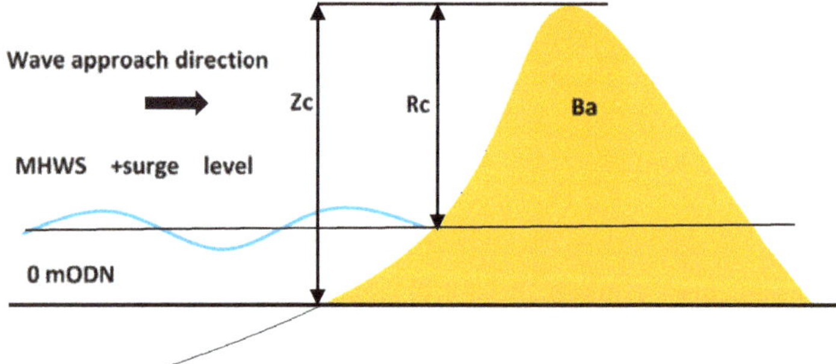

Figure 6. Schematisation of barrier geometry. B_a = pre-storm barrier cross sectional area above Surge level mODN + MHWS, R_c = initial barrier freeboard and Z_c = initial barrier crest height.

Figure 7 compares numerically simulated barrier inertia with Bradbury et al. [6] barrier inertia threshold at HS1 and HS2 from the 5 extreme storm conditions given in Table 3. At HS1, only the most severe storm event, 1:100, resulted in small amount of wave overtopping. In contrast, all storm events other than the 1:1 storm resulted in some overwashing or overtopping at HS2. Both results are in good agreement with the BIM, as shown in Figure 7, providing independent confirmation that the HCS model provides a reliable assessment of whether overwashing will occur during a storm.

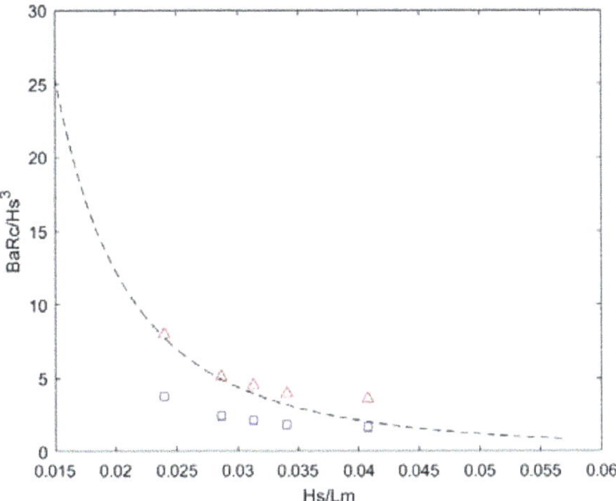

Figure 7. Comparison of Bradbury [9] and Bradbury et al. [6] Barrier Inertia Model and XBeach simulations. Black broken line-overwash threshold given by barrier inertia threshold of Bradbury [9]. Red triangles and blue squares are XBeach simulations at HS1 to HS2, respectively.

3.3. Simulation of Gravel Beach Profile Change

In order to develop a simple parametric model to estimate gravel barrier profile change of a wide variety of gravel barrier beaches under storm conditions, the validated XBeach model was used to generate 880 barrier overwash and volume realisations. A range of barrier cross-sections and storm conditions were used. Synthetic storm conditions were developed following a statistical analysis of long-term wave measurements of the Milford wave buoy, described in Section 2. Fifteen statistically significant storm wave heights with varying return periods between 1:1 and 1:100 years, determined by Bradbury et al. [6], in which a Weibull distribution was fitted to 19 years of wave buoy data. JONSWAP unimodal spectrum was used to generate storm wave conditions. The duration of storms was kept constant at 20 h which is the mean storm duration calculated using storms derived from the wave data measured at Milford wave buoy. Storm threshold wave height of 2.74 m, defined by the CCO was used to isolate storms from the wave measurements [46].

Six extreme sea levels (ESL) levels with return periods 1:1, 1:10, 1:20, 1:50, 1:100 and 1:200 were derived from Coastal flood boundary conditions for UK mainland and islands (DEFRA 2018) [47]. Those were combined with tide data to determine total water levels. It was assumed that the storm peak and the peak storm surge occur at the highest tide. Five different barrier beach cross sections from HCS with varying shapes and crest elevations were selected. The storm conditions, water levels and profile shapes were then combined to generate 880 physically plausible realisations of input conditions (Table 4) to drive the XBeach model to simulate profile change and overwash. 36 realisations of bimodal storms with six different swell percentages were also generated using the data given in Table 4. Bimodal spectra to derive offshore storm boundary conditions to the XBeach model, were determined using the approach given by Polidoro and Dornbusch [48]. The significant wave height Hs for bimodal waves was determined using:

$$H_s = \sqrt{H_{wind}^2 + H_{swell}^2} \tag{8}$$

where H_{wind} is the significant wave height of the wind wave component and H_{swell} is the significant wave height of the swell wave component.

Table 4. Significant wave heights, peak periods and storm surges with a range of return periods used to generate synthetic unimodal and bimodal storm events.

Unimodal Cases			Bimodal Cases		
H_s (m)	T_p (s)	MHWS + Surge above ODN (ESL) (m)	Hs (Bimodal) (m)	MHWS + Surge above ODN (ESL) (m)	Swell Percentage
2.75, 2.94, 3.13, 3.31, 3.50, 3.64, 3.76, 3.87, 3.99, 4.11, 4.22, 4.34, 4.46, 4.57, 4.69, 4.75	8.0, 9.0, 10.0, 11.5	1.52, 1.80, 1.96, 2.07, 2.20, 2.40	2.75, 3.64, 4.10, 4.75	1.52, 1.80, 1.96, 2.07, 2.20, 2.40	10, 25, 35, 40, 50, 75

4. Results and Discussion

XBeach model simulations of gravel barrier overwashing and volume change were carried out for both unimodal and bimodal conditions explained in Section 3. Several modes of barrier response were observed (Figure 8), which have been categorised using the conceptual barrier response model of McCall et al. [20] (please note only the cross section change above + 0 mODN is shown):

(1) *Beach face erosion*—For wave heights above the storm threshold height (2.74 m) combined with small peak wave periods ($T_p < 11.5$ s), where storm surge did not significantly reduce barrier freeboard (R_c), wave run up was confined to the swash zone. This resulted in sediment transported predominately offshore hence eroding the beach face (Figure 8A). Similar observations were found in Sallenger [49];

(2) *Crest accumulation*—In cross sections with small freeboard and/or barrier area, crest build-up due to overtopping was observed under moderate storm conditions. A similar process was observed in profiles with larger freeboard under higher energy storm conditions. In both cases sediment was typically transported up the beach face due to an increased run-up and deposited on the barrier crest (Figure 8B). This process reduced the width of the barrier. Gravel sediment transport on beach face is well described by [15]. Bradbury and Powel [5] state that crest accumulation can occur when barrier is rolled over and, a new crest may form at a higher elevation and behind the original location of crest if there is sufficient sediment in the system. However, our results did not show evidence of this process;

(3) *Crest lowering*—When energetic storm wave conditions (particularly with large wave periods) coincided with large surges, wave run-up exceeded the barrier freeboard and sediment was overwashed and deposited at the back of the barrier. As a result, the barrier crest was lowered and the width increased (Figure 8C). It was also observed that waves with low steepness increased overtopping. There were several cases where crest was lowered through avalanching of the barrier beach face;

(4) *Barrier Overwash*—Once a barrier had started experiencing overwashing, the general trend was that an increase in surge, H_s and T_p resulted in more overwash, leading to more sediment being deposited further behind the barrier (Figure 8D). The larger values of T_p resulted in sediment deposited further away from the back of the barrier. If ESL is significantly large, overwashing occurred even during low wave energy conditions. When the most energetic storms combined with largest storm surges overwash sediment was deposited far behind the barrier thus losing sediment from the active barrier morphodynamic system. As a result, the barrier may be more vulnerable to future wave attack, with long term effect being landward translation of the barrier, unless coastal management interventions take place.

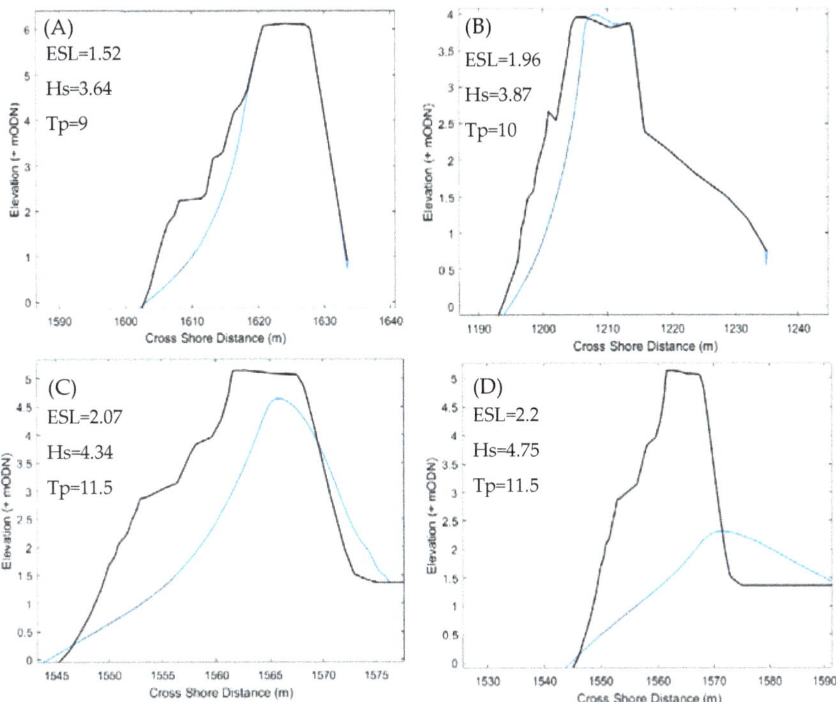

Figure 8. Observed morphodynamic responses of the barrier beach to storm events, from the simulated results. Black line-Pre-storm, Blue line-Post-storm. (**A**) Beach face erosion; (**B**) Crest accumulation; (**C**) Crest lowering; and (**D**) Barrier overwash.

Other observations of barrier morphodynamic response were as follows: Barrier overtopping and/or overwashing was unlikely when both B_a and R_c are large; when B_a is large but R_c is small, the barrier was more prone to overwashing. Cases with a small B_a coupled with small R_c, experienced similar overwashing to that with large B_a and small R_c. This highlights the importance of R_c as an essential defence against extreme run-up that can occur during high surges. B_a appears to be of significance in controlling how soon the barrier becomes susceptible to overwashing events whereas smaller values of B_a can result in overwashing occurring earlier.

Bimodal impacts on HCS were found to be increase in barrier volume change and also, more importantly, altering the mode of barrier response. Where a barrier cross section was not previously susceptible to overwashing under unimodal conditions, bimodal conditions with high swell percentages with same energy were capable of producing severe overwashing events on the same barrier cross section.

To capture the change in barrier volume, an approach similar to Bradbury [9] was used by coupling key hydrodynamic and geometric variables. The extreme sea level, H_s and T_p were taken as the key hydrodynamic parameters while B_a, R_c and Z_c (Figure 6) were taken as the key barrier geometric parameters. In order to better categorise the observed barrier responses using the controlling key parameters, a series of non-dimensional parameters were derived using parametric testing, which includes the Barrier inertia parameter of Bradbury [9] (hereafter known as B_i) given in Equation (9):

$$B_i = \frac{R_c B_a}{H_s^3} \tag{9}$$

B_i has already been used to at numerous previous occasions to detect if barrier overwash is likely to occur on gravel beaches, as mentioned in Section 4. It includes the importance of surge level on barrier response, making it appropriate for use also in this study.

In Figure 9, the simulated non-dimensional barrier volume change above 0 m ODN per metre length of the barrier multiplied by wave steepness is shown against the square root of the barrier inertia parameter (B_i). The results show a clear trend where smaller B_i will give rise to larger barrier volume change as expected. The trendline derived using regression analysis and the 95% confidence intervals are also shown. The regression curve with $r^2 = 0.82$ is given by the equation:

$$\frac{\Delta Vol}{Z_c^2}\frac{H_s}{L_p} = 0.02655\left(\frac{R_c B_a}{H_s^3}\right)^{-1.06} + 0.01 \tag{10}$$

where ΔVol is volume change per metre width of the barrier and L_p is deep water wave length corresponding to peak wave period T_p.

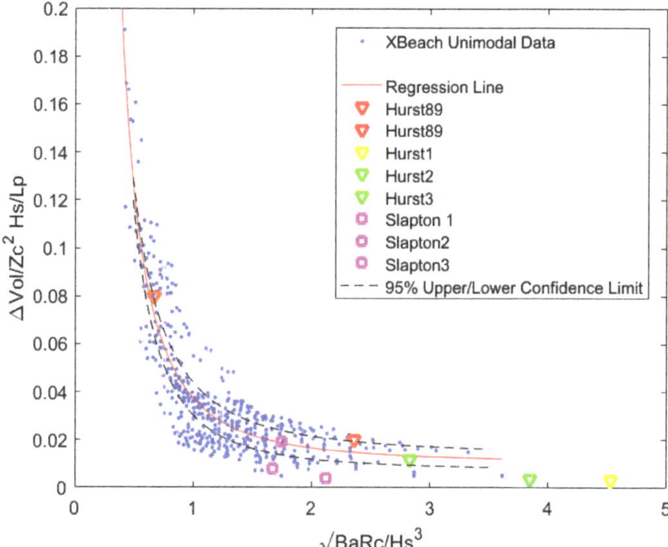

Figure 9. Non-dimensional barrier volume change above 0 m ODN per metre length of the barrier a during storm against the square root of the barrier inertia parameter (B_i). The exponential regression trendline is given by the red curve. 95% confidence limits are shown by the broken black curve. The regression curve validation data at HCS and Slapton Barrier beach are shown in green, yellow and red triangles and purple squares.

The valid range of the regression equation should not exceed, which are the physical limits of conditions modelled in XBeachX:

$$0.27 < \frac{R_c B_a}{H_s^3} < 13$$

and:

$$0.0133 < \frac{H_s}{L_p} < 0.047$$

It may be useful if the regression curve given in Equation (10) can be used as a predictor to estimate change in barrier volume during storms, which can serve as a simple parametric model.

To check the validity of Equation (10) for conditions outside those used for numerical simulations, barrier volume change measured at several cross sections of HCS and the Slapton barrier beach located in the south-west of the UK during a range of storms were used (Table 5). The historic pre- and post-storm barrier cross sections and, storm wave and water level data are provided by the Channel Coastal Observatory of the UK, Bradbury [9], Bradbury et al. [6], McCall et al. [38] and Chadwick et al. [14].

Table 5. Volume change measured at cross sections along HCS and Slapton beach during storm events, used to validate Equation (10).

Validation Case	H_s (m)	T_p (s)	Water Level above ODN (m) (ESL)
Hurst 89	2.9	10.96	0.87
Hurst 89	2.9	10.96	0.87
Hurst 1	3	12.6	1
Hurst 2	3.95	12.3	1.27
Hurst3	3.95	12.3	1.27
Slapton 1	4.87	8.3	1.905
Slapton 2	4.87	8.3	1.905
Slapton 3	4.87	8.3	1.905

The results reveal that two validation cases are outside the validity of Equation (10) $\left[\frac{R_c B_a}{H_s^3} > 13 \right]$. However, those cases resulted in very small barrier volume change and therefore, not significant. Four validation cases fell within the 95% confidence limits showing good degree of accuracy of the model while two cases were outside 95% confidence limit (Figure 9). Although further validation is necessary before the Equation (10) to be widely used as a parametric predictive model, these results give reasonable confidence to use it to estimate the changes in gravel barrier volume during storm events.

4.1. Simulated Bimodal Conditions Compared to Simulated Unimodal Conditions

Although the parametric model given in Equation (10) is derived based on numerically simulated barrier volume change during unimodal storm conditions, it is well known that the south-west of the UK is subjected to frequent storms with bimodal characteristics as a results of swell-dominated waves reaching from the Atlantic as mentioned in Section 2 [25–27]. To examine the impacts of bimodal storm conditions on gravel barriers, the model was used to simulate barrier volume change and overwash from bimodal storms explained in Section 3. In Figure 10 the non-dimensional barrier volume change from bimodal waves per metre length of the barrier are shown and compared with unimodal results. It can be seen that during bimodal storms with swell percentage greater than 40–50%, the non-dimensional barrier volume change for a given barrier inertia is larger than that from their unimodal counterparts, especially at low values of barrier inertia. There is a noticeable increase in barrier volume change from storms with 50% and 75% swell conditions. On the other hand, if the swell percentage is less than 30% non-dimensional barrier volume change is less than that from the unimodal cases for a given barrier inertia.

To examine the impact of wave bimodality on barrier response in more details, barrier volume change and overwash volume from storms under bimodal conditions at each individual cross section was investigated in isolation and compared with unimodal results. Figure 11 shows the results for cross section HS2, seen in Figure 4. Figure 11 clearly shows that the increasing swell wave component leads to greater barrier volume change and larger overwashing volumes and that when swell percentage is greater than 40% the barrier volume change and overwash volume is greater than that from the unimodal waves, for a given freeboard.

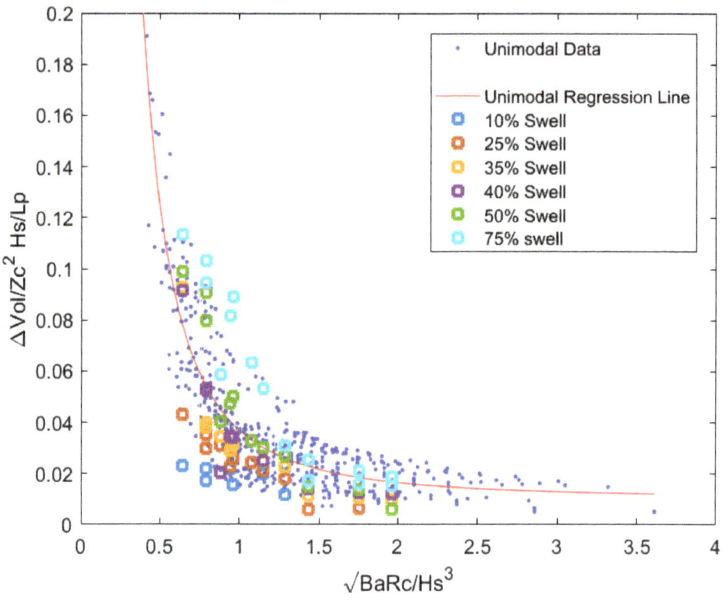

Figure 10. Comparison of unimodal and bimodal non-dimensional barrier volume change under different swell percentages. The red curve is the parametric model given by Equation (10).

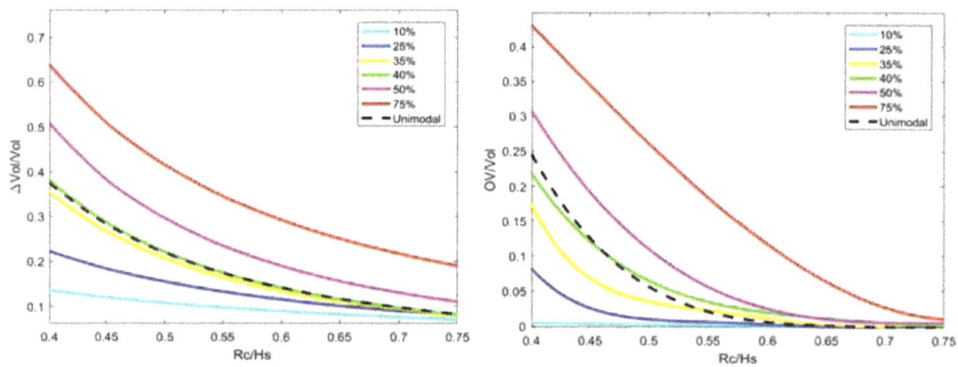

Figure 11. The effect of swell percentage of bimodal storm waves on barrier volume change (**left**) and overwash volume (**right**) for a single cross section, HS2. Δvol = barrier volume change per metre length and Vol = pre-storm barrier volume/metre length; OV = overwash volume.

4.2. Simulation of Gravel Beach Sediment Overwash Volume

Gravel barrier beach overwashing is an important process which contributes to back barrier flooding as well as barrier response to storms. Therefore, the relationship between the overwashing volume and the barrier inertia parameter B_i was also investigated. Figure 12 gives the numerically simulated non-dimensional overwash volume from both unimodal and bimodal storm conditions. Although there is significant data scatter, an overall trend of overwash volume reduction with increase in barrier inertia can be seen. Also, similar to barrier volume change, overwash volumes from bimodal storms with higher swell percentages are significantly larger than that from unimodal storms for the same barrier inertia. During extreme cases where swell percentage is 75%, complete inundation of the barrier has taken place and a significant volume of sediment has been removed from the active barrier beach systems and transported further away from the back barrier. An event similar to this has been recorded at HCS where the barrier has breached following a storm in 2005 [25]. On the other hand, storms with less than 25% swell component gave rise to notably low overwash volumes.

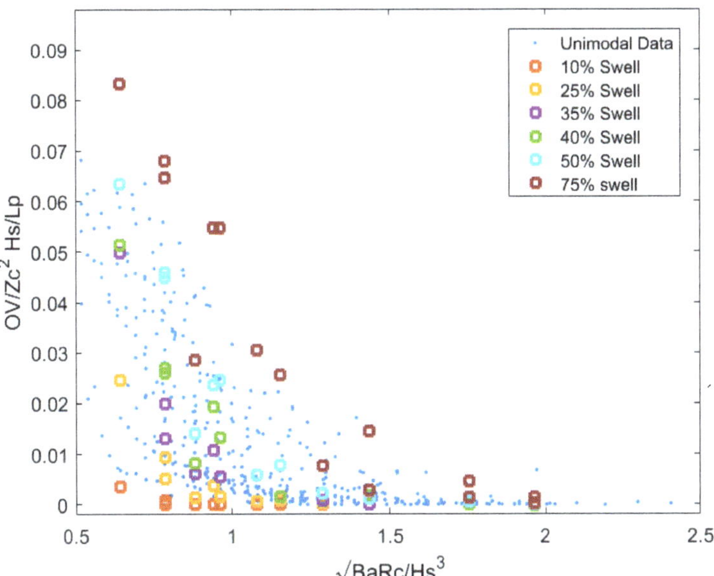

Figure 12. A comparison of simulated barrier overwash volumes from unimodal storms and bimodal storms with varying swell percentages.

A probable reason for 'positive' and 'negative' swell effects on the barrier volume change and overwashing may be explained by distinctly different behaviours of wind and swell waves. Short period wind waves dissipate further away from shore due to shoaling and steepening of the wave. This may limit the wave energy reaching the barrier beach face and wave run up. On the other hand, longer period swell waves, with their low wave steepness, can dominate the surf zone and propagate closer to the beach face without dissipation due to breaking [41,50]. If the swell percentage is small, wind waves dominate the surf, swash and runup while the contribution from swell waves may be small. For sea states with larger swell components, undissipated swell waves drive large runup and overtopping/overwashing. Polidoro et al. [13] observed that a swell percentage greater than 20% could have a larger impact on the elevation of the beach crest than the horizontal displacement of the beach, which agrees with our results.

5. Conclusions

The paper uses an extensive set of numerically simulated beach volume change and overwash data to developed a simple parametric model to estimate gravel barrier change under unimodal storm conditions and to investigate the response of gravel barriers to bimodal storm wave conditions. Following conclusions were drawn from the results:

- The XBeach non hydrostatic model is capable of simulating barrier volume change and overwash volume. The model was able to capture swash dynamics, sediment movement, barrier face erosion, crest build-up and back barrier sediment accumulation correctly, which was essential to the parametric model development in this study;
- The gravel barrier volume change during a collection of storms calculated using the parametric model were in good agreement with volume change measured in the field.

This proves that the model will be a useful tool to estimate barrier volume change during storms, which can be taken as first estimates for coastal management purposes.

- Bimodal storm waves with large swell percentages (>50%) lead to greater barrier volume change and larger overwash volumes than their unimodal counterparts. This can be explained by the action of low steepness, high energy wave propagation on the slope of the barrier giving rise to higher runup and sediment movement on the face of the barrier.
- Following limitations of the approach are noted: Further validation of the parametric model is necessary to extend its application to a wide range of gravel barriers; the numerical simulations were carried out in 1D where the impacts of longshore transport were not taken into consideration; sea level change due to global warming is not considered in the simulations; and the parametric model may underestimate barrier volume change from bimodal storm conditions. Further studies will be carried out in the future to address those limitations.

Author Contributions: Conceptualization, H.K. and D.P.; Methodology, K.I., H.K. and D.P.; Validation K.I.; Formal Analysis, K.I. and H.K.; Investigation, K.I.; Resources, H.K., D.P.; Writing K.I. and H.K.; Writing—Review & Editing, H.K., D.E.R. and D.P.; Supervision, H.K. and D.P.; Project Administration, H.K.; Funding Acquisition, H.K. and D.P. All authors have read and agreed to the published version of the manuscript.

Funding: This research was funded by the Knowledge Economy Skills Scholarships (KESS II) which is a pan-Wales higher level skills initiative led by Bangor University on behalf of the Higher Education sector in Wales, UK and JBA Consulting.

Data Availability Statement: All results presented in this paper can be received upon request to the corresponding author. Data used can be accesses via UK Channel Coastal Observatory.

Acknowledgments: K.I. acknowledges the Knowledge Economy Skills Scholarships (KESS II) which is a pan-Wales higher level skills initiative led by Bangor University on behalf of the Higher Education sector in Wales, UK. The project was part funded by the Welsh Government's European Social Fund (ESF) programme for East Wales and JBA Consulting enabling the pursuit of his MSc degree at Swansea University, UK. Channel Coastal Observatory is acknowledged for providing numerous data used in this study.

Conflicts of Interest: The authors declare no conflict of interest.

References

1. Carter, R.W.G.; Orford, J.D.; Lauderdale, F.; Carter, R.W.G. The Morphodynamics of Coarse Clastic Beaches and Barriers: A Short- and Long-term Perspective. *J. Coast. Res.* **1993**, *15*, 158–179.
2. Aminti, P.; Cipriani, L.E.; Pranzini, E. 'Back to the Beach': Converting Seawalls into Gravel Beaches. In *Soft Shore Protection. Coastal Systems and Continental Margins*; Goudas, C., Katsiaris, G., May, V., Karambas, T., Eds.; Springer: Dordrecht, The Netherlands, 2003; Volume 7. [CrossRef]
3. Buscombe, D.; Masselinki, G. Concepts in gravel beach dynamics. *Earth-Sci. Rev.* **2006**, *79*, 33–52. [CrossRef]
4. Almeida, L.P.; Masselink, G.; Russell, P.E.; Davidson, M.A. Observations of gravel beach dynamics during high energy wave conditions using a laser scanner. *Geomorphology* **2015**, *228*, 15–27. [CrossRef]
5. Bradbury, A.P.; Powell, K.A. The Short Term Profile Response of Shingle Spits to Storm Wave Action. In Proceedings of the 23rd International Conference on Coastal Engineering, Venice, Italy, 4–9 October 1992; Volume 3, pp. 2694–2707.
6. Bradbury, A.P.; Cope, S.N.; Prouty, D.B. Predicting the response of shingle barrier beaches under extreme wave and water level conditions in southern England. In *Coastal Dynamics 2005*; American Society of Civil Engineers: Reston, VA, USA, 2005.
7. Powell, K.A. The dynamic response of shingle beaches to random waves. In Proceedings of the 21st International Conference on Coastal Engineering, Malaga, Spain, 20–25 June 1988; pp. 1763–1773.
8. Powell, K.A. *Predicting Short-Term Profile Response for Shingle Beaches*; Hydraulics Research Report SR219; HR Wallingford: Wallingford, UK, 1990.
9. Bradbury, A.P. Predicting breaching of shingle barrier beaches-recent advances to aid beach management. In Proceedings of the 35th Annual MAFF Conference of River and Coastal Engineers, London, UK, July 2000; Available online: https://www.semanticscholar.org/paper/PREDICTING-BREACHING-OF-SHINGLE-BARRIER-ADVANCES-TO-Bradbury/ee91f390d2c45f6a0f99b31e5fb1e858213b87f3 (accessed on 17 December 2019).

10. de San Román-Blanco, B.L.; Coates, T.T.; Holmes, P.; Chadwick, A.J.; Bradbury, A.; Baldock, T.E.; Pedrozo-Acuña, A.; Lawrence, J.; Grüne, J. Large scale experiments on gravel and mixed beaches: Experimental procedure, data documentation and initial results. *Coast. Eng.* **2006**, *53*, 349–362. [CrossRef]

11. Van Rijn, L.C.; Sutherland, J.; Rosati, J.D.; Wang, P.; Roberts, T.M. Erosion of gravel barriers and beaches. In Proceedings of the Coastal Sediments 2011, Miami, FL, USA, 2–6 May 2011.

12. Matias, A.; Blenkinsopp, C.E.; Masselink, G. Detailed investigation of overwash on a gravel barrier. *Mar. Geol.* **2014**, *350*, 27–38. [CrossRef]

13. Polidoro, A.; Pullen, T.; Eade, J.; Mason, T.; Blanco, B.; Wyncoll, D. Gravel beach profile response allowing for bimodal sea states. *Proc. Inst. Civ. Eng. Marit. Eng.* **2018**, *171*, 145–166. [CrossRef]

14. Chadwick, A.J.; Karunarathna, H.; Gehrels, W.R.; Massey, A.C.; O'Brien, D.; Dales, D. A new analysis of the Slapton barrier beach system, UK. *Proc. Inst. Civ. Eng. Marit. Eng.* **2005**, *158*, 147–161. [CrossRef]

15. Austin, M.J.; Buscombe, D. Morphological change and sediment dynamics of the beach step on a macrotidal gravel beach. *Mar. Geol.* **2008**, *249*, 167–183. [CrossRef]

16. de Alegria-Arzaburu, A.R.; Masselink, G. Storm response and beach rotation on a gravel beach, Slapton Sands, UK. *Mar. Geol.* **2010**, *278*, 77–99. [CrossRef]

17. Williams, J.J.; De Alegria-Arzaburu, A.R.; McCall, R.T.; Van Dongeren, A.; Van Dongeren, A. Modelling gravel barrier profile response to combined waves and tides using XBeach: Laboratory and field results. *Coast. Eng.* **2012**, *63*, 62–80. [CrossRef]

18. Roelvink, D.; Reniers, A.; Van Dongeren, A.; van Thiel de Vries, J.; McCall, R.; Lescinski, J. Modelling storm impacts on beaches, dunes and barrier islands. *Coast. Eng.* **2009**, *56*, 1133–1152. [CrossRef]

19. Jamal, M.H.; Simmonds, D.J.; Magar, V. Modelling gravel beach dynamics with XBeach. *Coast. Eng.* **2014**, *89*, 20–29. [CrossRef]

20. McCall, R.T.; Masselink, G.; Poate, T.G.; Roelvink, D.; Almeida, L.P. Modelling the morphodynamics of gravel beaches during storms with XBeach-G. *Coast. Eng.* **2015**, *103*, 52–66. [CrossRef]

21. Gharagozlou, A.; Dietrich, J.C.; Karanci, A.; Luettich, R.A.; Overton, M.F. Storm-driven erosion and inundation of barrier islands from dune-to region-scales. *Coast. Eng.* **2020**, *158*, 103674. [CrossRef]

22. Phillips, B.T.; Brown, J.M.; Plater, A.J. Modeling Impact of Intertidal Foreshore Evolution on Gravel Barrier Erosion and Wave Runup with Xbeach-X. *J. Mar. Sci. Eng.* **2020**, *8*, 914. [CrossRef]

23. Roelvink, D.; McCall, R.; Mehvar, S.; Nederhoff, K.; Dastgheib, A. Improving predictions of swash dynamics in XBeach: The role of groupiness and incident-band runup. *Coast. Eng.* **2018**, *134*, 103–123. [CrossRef]

24. Poate, T.G.; McCall, R.T.; Masselink, G. A new parameterisation for runup on gravel beaches. *Coast. Eng.* **2016**, *117*, 176–190. [CrossRef]

25. Bradbury, A.P.; Mason, T.E.; Poate, T. Implications of the spectral shape of wave conditions for engineering design and coastal hazard assessment—Evidence from the English Channel. In Proceedings of the 10th International Workshop on Wave Hindcasting and Forecasting and Coastal Hazard Symposium, North Shore, Oahu, HI, USA, 11–16 November 2007; pp. 1–17.

26. Thompson, D.A.; Karunarathna, H.; Reeve, D.E. An Analysis of Swell and Bimodality around the South and South-west coastline of England. *Nat. Hazards Earth Syst. Sci. Discuss.* **2018**. Available online: https://nhess.copernicus.org/preprints/nhess-2018-117/ (accessed on 17 December 2019). [CrossRef]

27. Nicholls, R.; Webber, N. Characteristics of Shingle Beaches with Reference to Christchurch Bay, S. England. In Proceedings of the 21st International Conference on Coastal Engineering, Malaga, Spain, 20–25 June 1988; pp. 1922–1936. [CrossRef]

28. British Oceanography Data Centre BODC. Available online: https://www.bodc.ac.uk/data/hosted_data_sys-tems/sea_level/uk_tide_gauge_network/ (accessed on 18 January 2020).

29. Bradbury, A.P.; Kidd, R. Hurst spit stabilisation scheme. Design and construction of beach recharge. In Proceedings of the 33rd MAFF Conference of River and Coastal Engineers, Keele, UK, 1–3 July 1998; Volume 1. No. 1.

30. Nicholls, R. The Stability of Shingle Beaches in the Eastern Half of Christchurch Bay. Unpublished Ph.D. Thesis, Department of Civil Engineering, University of Southampton, Southampton, UK, 1985; 468p.

31. Bradbury, A.P. Evaluation of coastal process impacts arising from nearshore aggregate dredging for beach re-charge—Shingles Banks, Christchurch Bay. In Proceedings of the International Conference on Coastal Manage-ment 2003, Brighton, UK, 15–17 October 2003; pp. 1–15.

32. Channel Coastal Observatory (CCO). Available online: https://www.channelcoast.org/ (accessed on 18 January 2020).

33. Williams, J.J.; Esteves, L.S.; Rochford, L.A. Modelling storm responses on a high-energy coastline with XBeach. *Model. Earth Syst. Environ.* **2015**, *1*, 1–14. [CrossRef]

34. Holthuijsen, L.H.; Booij, N.; Herbers, T.H.C. A prediction model for stationary, short-crested waves in shallow water with ambient currents. *Coast. Eng.* **1989**, *13*, 23–54. [CrossRef]

35. Smit, P.B.; Roelvink, J.A.; van Thiel de Vries, J.S.M.; McCall, R.T.; van Dongeren, A.R.; Zwinkels, J.R. *XBeach Manual: Non-Hydrostatic Model. Validation, Verification and Model Description*; Delft University of Technology: Delft, The Netherlands, 2010.

36. Smit, P.; Zijlema, M.; Stelling, G. Depth-induced wave breaking in a non-hydrostatic, near-shore wave model. *Coast. Eng.* **2013**, *76*, 1–16. [CrossRef]

37. Zijlema, M.; Stelling, G.; Smit, P. SWASH: An operational public domain code for simulating wave fields and rapidly varied flows in coastal waters. *Coast. Eng.* **2011**, *58*, 992–1012. [CrossRef]

38. McCall, R.; Masselink, G.; Poate, T.; Bradbury, A.; Russell, P.; Davidson, M. Predicting overwash on gravel barri-ers. *J. Coast. Res.* **2013**, 1473–1478. [CrossRef]
39. McCall, R.T.; Masselink, G.; Poate, T.G.; Roelvink, J.A.; Almeida, L.P.; Davidson, M.; Russell, P.E. Modelling storm hydrodynamics on gravel beaches with XBeach-G. *Coast. Eng.* **2014**, *91*, 231–250. [CrossRef]
40. Van Rijn, L.C.; Walstra, D.J.R.; Grasmeijer, B.; Sutherland, J.; Pan, S.; Sierra, J.P. The predictability of cross-shore bed evolution of sandy beaches at the time scale of storms and seasons using process-based Profile models. *Coast. Eng.* **2003**, *47*, 295–327. [CrossRef]
41. Masselink, G.; McCall, R.; Poate, T.; Van Geer, P. Modelling storm response on gravel beaches using XBeach-G. *Proc. Inst. Civ. Eng. Marit. Eng.* **2014**, *167*, 173–191. [CrossRef]
42. Orimoloye, S.; Karunarathna, H.; Reeve, D.E. Effects of swell on wave height distribution of energy-conserved bimodal seas. *J. Mar. Sci. Eng.* **2019**, *7*, 79. [CrossRef]
43. Soulsby, R.L.; Whitehouse, R.J.S. Threshold of sediment motion in coastal environments. In *Pacific Coasts and Ports' 97: Proceedings of the 13th Australasian Coastal and Ocean Engineering Conference and the 6th Australasian Port and Harbour Conference*; Centre for Advanced Engineering, University of Canterbury: Christchurch, New Zealand, 1997; Volume 1.
44. Van Rijn, L.C. Unified View of Sediment Transport by Currents and Waves. III: Graded Beds. *J. Hydraul. Eng.* **2007**, *133*, 761–775. [CrossRef]
45. Van Rijn, L.C. *Principles of Sediment Transport in Rivers, Estuaries and Coastal Seas*; Aqua Publications: Amsterdam, The Netherlands, 1993; Volume 1006.
46. XBeach Manual, Model Description and Reference Guide to Functionalities, Deltares. 2015. Available online: https://xbeach.readthedocs.io/en/latest/xbeach_manual (accessed on 17 December 2019).
47. Department for Environment, Food and Rural Affairs (DEFRA) DE-FRA. Available online: https://www.gov.uk/government/organisations/department-for-environment-food-rural-affairs (accessed on 17 December 2019).
48. Polidoro, A.; Dornbusch, U. *Wave Run-Up on Shingle Beaches—A New Method*; HR Wallingford Technical Report CAS0942-RT001-R05-00; HR Wallingford: Wallingford, UK, 2014.
49. Sallenger, A.H., Jr. Storm impact scale for barrier islands. *J. Coast. Res.* **2000**, *16*, 890–895.
50. Almeida, L.P.; Masselink, G.; McCall, R.; Russell, P. Storm overwash of a gravel barrier: Field measurements and XBeach-G modelling. *Coast. Eng.* **2017**, *120*, 22–35. [CrossRef]

Journal of
*Marine Science
and Engineering*

Article

Rational Evaluation Methods of Topographical Change and Building Destruction in the Inundation Area by a Huge Tsunami

Sayed Masihullah Ahmadi [1,*], Yoshimichi Yamamoto [2] and Vu Thanh Ca [3]

[1] Graduate School of Science and Technology, Tokai University, Kanagawa 259-1292, Japan
[2] Department of Civil Engineering, Tokai University, Kanagawa 259-1292, Japan; yo-yamamo@tokai.ac.jp
[3] Department of Environment, Ha Noi University of Natural Resources and Environment, 41A Phu Dien, Bac Tu Liem, Hanoi, Vietnam; vtca@hunre.edu.vn
* Correspondence: 9ltad003@mail.u-tokai.ac.jp

Received: 4 August 2020; Accepted: 27 September 2020; Published: 7 October 2020

Abstract: In the case of huge tsunamis, such as the 2004 Great Indian Ocean Tsunami and 2011 Great East Japan Tsunami, the damage caused by ground scour is serious. Therefore, it is important to improve prediction models for the topographical change of huge tsunamis. For general models that predict topographical change, the flow velocity distribution of a flood region is calculated by a numerical model based on a nonlinear long wave theory, and the distribution of bed-load rates is calculated using this velocity distribution and an equation for evaluating bed-load rates. This bed-load rate equation usually has a coefficient that can be decided using verification simulations. For the purpose, Ribberink's formula has high reproducibility within an oscillating flow and was chosen by the authors. Ribberink's formula needs a bed-load transport coefficient that requires sufficient verification simulations, as it consumes plenty of time and money to decide its value. Therefore, the authors generated diagrams that can define the suitable bed-load coefficient simply using the data acquired from hydraulic experiments on a movable bed. Subsequently, for the verification purpose of the model, the authors performed reproduced simulations of topography changes caused by the 2011 Great East Japan Tsunami at some coasts in Northern Japan using suitable coefficients acquired from the generated diagrams. The results of the simulations were in an acceptable range. The authors presented the preliminary generated diagrams of the same methodology but with insubstantial experimental data at the time at the International Society of Offshore and Polar Engineers (ISOPE), (2018 and 2019). However, in this paper, an adequate amount of data was added to the developed diagrams based on many hydraulic experiments to further raise their reliability and their application extent. Furthermore, by reproducing the tsunami simulation on the Sendai Natori coast of Japan, the authors determined that the impact of total bed-load transport was much bigger than that of suspension loads. Besides, the simulation outputs revealed that the mitigation effect of the cemented sand and gravel (CSG) banks and artificial refuge hills reduced tsunami damage on Japan's Hamamatsu coast. Since a lot of buildings and structures in the inundation area can be destroyed by tsunamis, building destruction design was presented in this paper through an economy and simplified state. Using the proposed tsunami simulation model, we acquired the inundation depth at any specific time and location within the inundated area. Because the inundation breadth due to a huge tsunami can extend kilometers toward the inland area, the evaluation of building destruction is an important measure to consider. Therefore, the authors in this paper presented useful threshold diagrams to evaluate building destruction with an easy and cost-efficient state. The threshold diagrams of "width of a pillar" for buildings or "width of concrete block walls" not breaking to each inundation height were developed using the data of damages due to the 2011 Great East Japan Tsunami.

Keywords: tsunami inundation simulation; topography change due to a huge tsunami; wall/pillar resistance against tsunami; CSG dike

---

1. Introduction

Research for predicting coastal scour and erosion by waves or flow has been performed all over the world. In particular, there has been a series of research regarding the prediction of the coastal scour and erosion within a wide area due to tsunamis. Takahashi et al. (1993a, 1993b) [1,2] proposed a formula to estimate bed-load transport to predict scour due to tsunamis. They assumed that the bed-load transport was proportional to power n of the Shields number. However, they also stated that it was not possible to neglect the suspended load transport. Kobayashi et al. (1996) [3] proposed a formula to estimate bed-load transport with the power *1.5* of Shields number and found a good agreement with experimental results. Fujii et al. (1998) [4] and Takahashi et al. (2000) [5] considered suspended load entrainment and deposition that improved the prediction accuracy of the bottom topography change due to tsunamis. Nishihata et al. (2006) [6] considered both bed-load and suspended load when developing a numerical model to estimate sediment deposit flux in Sri Lanka's Kirinda fishing harbor. Kihara and Matsuyama (2007) [7] developed a three-dimensional (3D) model to compute bottom topography changes caused by tsunamis. Nakamura and Mizutani (2008) [8] proposed a formula to estimate bed-load and suspended-load transport that accounts for the fluctuation of underground stress. Thus, improvement in the prediction model's accuracy has been advanced. However, the development of a model by which many coastal engineers can easily predict practical accuracy is also desired. Therefore, Ca, Yamamoto, and Charusrojthanadech (2010) [9] presented a two-dimensional numerical simulation model that calculated inundation velocity, inundation depth, and topographical change on an inundation area. For hydraulic calculations, a continuity equation and two-dimensional nonlinear shallow water equations were used. To calculate topographical changes, this model used Ribberink's equation. This equation requires a bed-load transport coefficient that is usually decided after using verification simulations. To decrease the time and cost needed to decide the value of Ribberink's bed-load transport coefficient, Ahmadi, Yamamoto, and Hayakawa (2018 and 2019) [10,11] performed many hydraulic experiments and developed useful diagrams to obtain the bed-load coefficient by using the inverse analysis of the experiment results. However, at the time, the amount of experiment data was inadequate. Therefore, in this research, more experimental data was added into the developed diagrams to decide if the bed-load transport coefficient further raised reliability. By reproducing the tsunami simulation on Japan's Sendai Natori coast, the suspension load showed a smaller effect on topography change as the total bed-load. Moreover, as a mitigation effect to evaluate tsunami countermeasures, we executed prediction simulations for tsunami inundation and topographical change on Japan's Hamamatsu coast. Since the local government constructed coastal banks of 13 m in height made of cemented sand and gravel (CSG) and some evacuation soil mounds, we examined the effect that CSG banks and artificial refuge hills had on reducing tsunami damage.

As a final result, this paper presented a rational method for predicting building destruction caused by tsunamis. Previously, Yamamoto et al. (2006) [12] showed that the stress analysis using the gate-type Rahmen model could accurately predict whether each building was broken by a tsunami or not. Our proposed hydro-morphodynamics model made it possible to calculate inundation depth and inundation velocity at any desired point and time in a proposed area. Yamamoto, Nunthawath, and Nariyoshi (2011) [13], as well as Charusrojthanadech, Yamamoto, and Ca (2011, 2012) [14,15], showed that the combination of tsunami simulations and diagrams made by stress analysis using the gate-type Rahmen mode easily predicted whether buildings were broken by a tsunami. The formal evaluation method of building destruction due to tsunamis are usually costly and time-consuming, while methods using developed diagrams are economical and simple. The threshold diagrams in their papers were developed using damage surveys from Japan's 2004 Indian Ocean Tsunami and

1993 Hokkaido-Nansei-Oki Tsunami. Using these diagrams, we evaluated the limit width of a wall or pillar to be safe against a probable tsunami inundation depth. However, since the material quality of Japanese buildings broken by the 2011 Great East Japan Tsunami differed from the quality of buildings in Thailand and Sri Lanka broken by the 2004 Indian Ocean Tsunami, we developed new threshold diagrams to evaluate building destruction using destruction data of the 2011 Great East Japan Tsunami. As a result, we evaluated the limit width of a wall or pillars to be safe against tsunami inundation depth. The evaluation method using threshold diagrams was found to be easy, economical, and useful in countries like Japan.

2. Rational Method for Predicting Topographical Change by Tsunami

2.1. Existing Numerical Simulation Model

To predict topographical change caused by tsunamis, we used Ca et al.'s (2010) [9] numerical simulation model. The model was developed using the following detailed formulas.

2.1.1. Numerical Model for Fluid Motion

The numerical model used for tsunami simulations in this paper was based on a continuity equation of fluid (Equation (1)) and two-dimensional nonlinear long wave equations (Equations (2) and (3)). These governing equations were solved by finite difference methods using the Crank–Nicholson scheme.

$$\frac{\partial f_y q_x}{\partial x} + \frac{\partial f_x q_y}{\partial y} + \frac{\partial S\eta}{\partial t} = 0 \tag{1}$$

$$\frac{\partial q_x}{\partial t} + \frac{1}{S}\frac{\partial}{\partial x}\left(\frac{Sq_x^2}{d}\right) + \frac{1}{S}\frac{\partial}{\partial y}\left(\frac{Sq_x q_y}{d}\right) + gd\frac{\partial \eta}{\partial x} - \frac{1}{S}\frac{\partial}{\partial x}\left\{dv_t S\frac{\partial}{\partial x}\left(\frac{q_x}{d}\right)\right\} - \frac{1}{S}\frac{\partial}{\partial y}\left\{dv_t S\frac{\partial}{\partial y}\left(\frac{q_x}{d}\right)\right\} + \frac{f_c}{d^2}Qq_x = 0 \tag{2}$$

$$\frac{\partial q_y}{\partial t} + \frac{1}{S}\frac{\partial}{\partial x}\left(\frac{Sq_y q_x}{d}\right) + \frac{1}{S}\frac{\partial}{\partial y}\left(\frac{Sq_y^2}{d}\right) + gd\frac{\partial \eta}{\partial y} - \frac{1}{S}\frac{\partial}{\partial x}\left\{dv_t S\frac{\partial}{\partial x}\left(\frac{q_y}{d}\right)\right\} - \frac{1}{S}\frac{\partial}{\partial y}\left\{dv_t S\frac{\partial}{\partial y}\left(\frac{q_y}{d}\right)\right\} + \frac{f_c}{d^2}Qq_y = 0 \tag{3}$$

where q_x and q_y are the horizontal fluid fluxes in the x and y directions, respectively. η is the water surface elevation. f_x and f_y are the x and y direction ratios of the wet portion in a calculation mesh. S is the area ratio of the wet portion in a calculation mesh. d is the water depth (from the static water surface+ η). g is the gravitational acceleration. v_r is the eddy viscosity coefficient. f_c is the ground surface friction coefficient and $Q(= \sqrt{q_x^2 + q_y^2})$ is the compound value of q_x and q_y.

To calculate the eddy viscosity coefficient and the ground surface friction coefficient, the following equations were used:

$$v_t = \varepsilon\left[\left(\frac{\partial U}{\partial y}\right)^2 + \left(\frac{\partial V}{\partial x}\right)^2\right]^{\frac{1}{2}} \cdot d^2 \tag{4}$$

$$f_c = \frac{gn^2}{d^{\frac{1}{3}}}, \quad n^2 = n_0^2 + 0.020\frac{B_r}{100 - B_r}d^{\frac{4}{3}}, \quad n_0^2 = \frac{n_1^2 A_1 + n_2^2 A_2 + n_3^2 A_3}{A_1 + A_2 + A_3} \tag{5}$$

Here, U and V are the flow velocity in y and x directions. ε equals (0.1). n is the Manning's roughness coefficient. B_r is the building ratio (= the ratio of the area of all vertical objects like houses and to the mesh area). n_0 is the weighted average roughness coefficient of areas like farms, roads, and waste and wetlands, $A_{1,\,2,\,and3}$, with relative roughness coefficients of $n_{1,\,2,\,and3}$.

2.1.2. Numerical Model for Topographical Change

Topographical change based on sediment transport can be expressed using the continuity equation shown in Equation (5):

$$\frac{\partial \zeta}{\partial t} = -\frac{1}{1 - \varepsilon_s}\left(\frac{\partial q_{bx}}{\partial x} + \frac{\partial q_{by}}{\partial y} - C_s + C_{ut}\right) \tag{6}$$

where ζ is the ground surface elevation, q_{bx} and q_{by} are the bed-load rate per unit width in x and y directions, respectively, C_s is the deposition rate of the suspended load, C_{ut} is the entrainment rate of the suspended load from the bed, and ε_s is the porosity of the sediment.

Modeling of q_x and q_y on Bed-load Transport

To evaluate the bed-load rate, Ribberink's formula (1998) [16], shown in Equation (6), was used. Yokoyama et al. (2002) [17] performed many flow and wave scouring calculations using indoor and outdoor data on sand and gravel, with diameter ranges of 0.2–10 mm. They found that accurate results could be obtained using this formula.

$$q_{bi} = \begin{cases} C_b\big[|\theta_s(t)| - \theta_{sc}\big]^{1.65} \frac{\theta_s(t)}{|\theta_s(t)|} \sqrt{\Delta g(D_{50})^3} & (\theta_s(t) \geq \theta_{sc}) \\ 0 & (\theta_s(t) < \theta_{sc}) \end{cases} \tag{7}$$

where q_{bi} is the bed-load transport rate per unit width in the i direction, C_b is the bed-load coefficient determined by verification simulations, $\theta_s(t)$ is the Shields parameter in the i direction, θ_{sc} is the critical Shields number, Δ is the relative density of the sand, g is the gravitational acceleration, and D_{50} is the median diameter of the sediment.

Modeling of C_s and C_{ut} on Suspended Load Transport

During tsunamis, it is necessary to consider the influence of suspended load transport. The deposition rate of the suspended load C_s and the entrainment rate from the bed C_{ut} was evaluated using Equation (7) based on the vertical distribution of suspended load concentration:

$$C_s = w_s C\left(\frac{w_s}{2}\right), \; C_{ut} = -v_t \frac{\partial C(z)}{\partial z}\Big|_{z=z_a}, \tag{8}$$

where w_s is the settling velocity of suspended particles. $C(z)$ is the suspended load concentration and v_t is the eddy viscosity.

According to Soulsby (1997) [18], when assuming a sheet flow condition for a whole area, the vertical distribution of sediment concentration is expressed as follows:

$$C(z) = C_a\left(\frac{z}{z_a}\right)^b, \; b = -\frac{w_s}{\kappa u_*} \tag{9}$$

where κ is Karman's constant (=0.4), u_* is the friction velocity, C_a and z_a are estimated using the equation of Zyserman and Fredsoe (1994) [19].

2.2. Rational Evaluation Method of Bed-Load Transport Rate

To calculate the bed-load transport using Ribberink's equation, we needed to perform many verification simulations to obtain a suitable bed-load transport coefficient. However, in this paper, after we performed many hydraulic experiments using inverse analysis, we developed useful diagrams to obtain the bed-load coefficients based on median grain size, uniformity coefficients, and dry soil density for a proposed area. The development of these diagrams saved time and money.

2.2.1. Hydraulic Model Experiment Method

Based on Iwanuma city's (2015) [20] tsunami protection plan in the Miyagi Prefecture, the model consisted of a typical beach profile, coastal dike, and a general land-side ground. They were set with a scale of 1/20 in the water channel, 0.8 m in height, 19.0 m in length, and 0.5 m in width (as shown in Figure 1). Some tsunami cases were reproduced by collecting water in a tank connected to the left end of the channel and then discharging the water into the channel. The experimental apparatus and measurement methods are illustrated below.

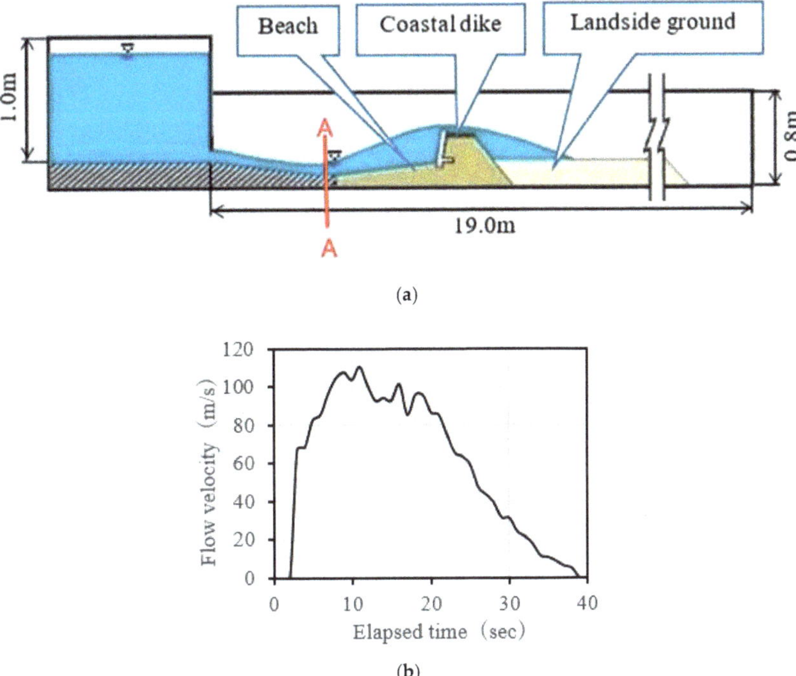

Figure 1. (a)Illustration of experimental apparatus for arising scour in the landside of the coastal dike. (b) Example of flow velocity on the crown part of the dike.

Four cases were tested with water heights of 45 cm, 60 cm, and 65 cm, which produced maximum flow velocities of 0.9 m/s, 1.05 m/s, and 1.1 m/s on the crown of the coastal dike, respectively. They are tabulated in Table 1. The number of all experiments was 25. The sediment of 0.2 mm in median grain size and 20.1 in uniformity coefficient was selected as the basic soil because it was common in Japan. In cases 3 and 4, grain sizes up to 10 mm and 30 mm, respectively, were used so that experimental results could be used in areas with larger grain sizes. Here, natural ones (a little round) were used in cases where median grain sizes were smaller than 20 mm. On the other hand, round gravel and crushed gravel were used in the case that the median grain size was 20 mm or 30 mm. Flow velocity on the crown of the dike was measured with an electromagnetic flow velocity meter of disk type and flow velocity in the scour area on the land side of the dike, which were measured using a propeller meter. Then, since the sidewalls of the water channel are made of transparent acrylic boards stiffened with angle irons, the topographical change of the ground was observed from the side by using a video camera, as shown in Figure 2. Moreover, the experiments were performed by a unit that was two times the same case.

Table 1. Parameters of scouring experiments on the dike's land side.

Case	1			2		3		4	
Max. Velocity on the Crown (m/s)	0.9 * (0.45)			1.1–1.25 * (0.65)		1.05–1.15 * (0.60)		1.0 * (0.45)	
Type of the sediment	Clay	Sand	Soil	Sand	Soil	Sand·Gravel	Soil	Gravel	Soil **
Median grain size (mm)	0.005	0.2	0.2	0.2	0.2	0.1, 0.2, 10	0.2	1.5, 3.4, 5.0, 20, 30	0.2
Uniformity coefficient	–	1.56	20.1	1.56	20.1	1.5–2	11	1.5–3	20.1
Dry density (g/cm³)				Around 1.5				Around 1.5	1.6–2.0

*: There was a difference in the flow velocity due to the effect of slightly different embankment heights depending on the implementation time. **: The concentrations of suspension load rates are around 10% and less than it.

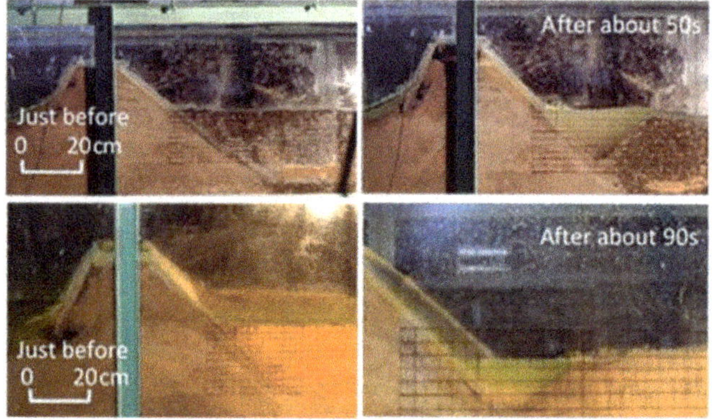

Figure 2. Experimental situation of scour in the dike's land side.

The flow velocity and the scour depth of the model experiments were converted to values in the field using Froude's similarity law. Moreover, "a field friction coefficient/a model friction coefficient = the 3/4th–1st power of the model scale" was obtained from existing experimental data, and this condition was substituted to Shields' similarity law, i.e., "a model grain size/a field grain size = the 1/4th–0th power of the model scale". Therefore, we considered the model grain size in our experiments as the same as the field grain size. Indeed, the maximum velocities near the coastal dike in Iwanuma City were 4–7 m/s. On the other hand, the maximum velocities of the model were $\div \sqrt{1/20} = 4$–5 m/s. The maximum scours depths near the coastal dike were 3–6 m. On the other hand, the maximum scour depths of the model were $\div 1/20 = 3$–5 m.

The simulations used to create Figures 3–5 are explained by the following three steps:

1) We reproduced the topographical feature on the right of the section (A) of Figure 1a and inputted time series data of water level from the section (A) so that time-series data of the simulated flow velocity on the crown part of the coastal dike was in agreement with the measured data (refer to Figure 1b).

2) We setup the first approximate value of the bed-load rate coefficient C_b, performed topographical change simulation, and calculated the maximum scour depth, thus forming the average value of scouring width by the dike's landside. If these two calculated values were mostly in agreement with the measured values, we considered the value of assumed C_b to be a true value.

3) When these two calculated values were not in agreement with the measured values, we changed the value of C_b and repeated the topographical change simulation until these two calculated values agreed.

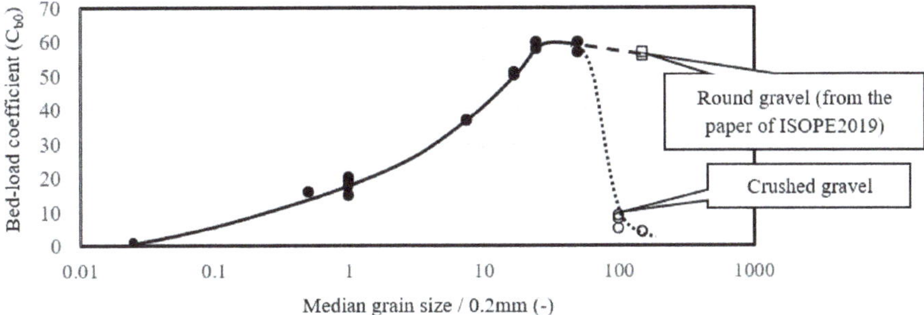

Figure 3. Influence of the median grain size to the bed-load coefficient. (Uniformity coefficients are 1.5–3; dry densities are around 1.5 g/cm^3).

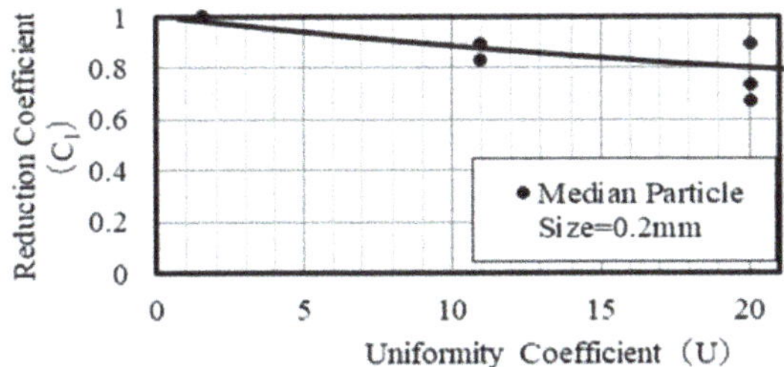

Figure 4. Influence of the uniformity coefficient to the bed-load reduction coefficient (the median grain size is 0.2 mm, dry densities are around 1.5 g/cm^3).

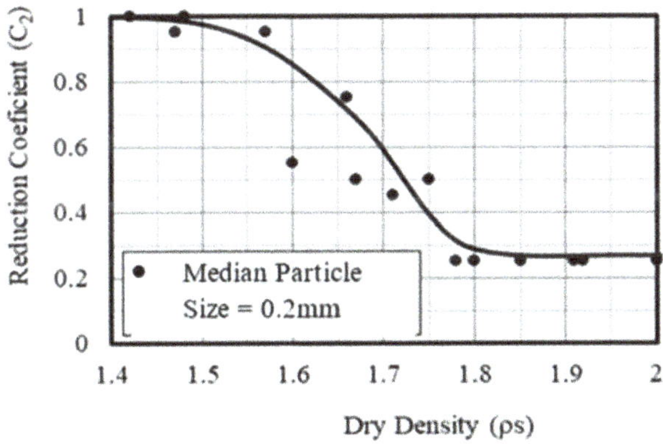

Figure 5. Influence of dry density to the bed-load reduction coefficient (the median grain size and the uniformity coefficient are 0.2 mm and 20.1 mm, respectively).

2.2.2. Bed-load Coefficient by Inverse Analysis

The bed-load coefficients sufficiently reproduced the scouring depth of the executed hydraulic experiments and were obtained using the numerical model of Ca et al. (2010) [9]. Furthermore, the influence of median grain sizes, uniformity coefficients, and dry densities on bed-load coefficients were also examined. As shown in Figure 3, when median grain size became large (5 mm), the bed-load coefficient "C_{b0}" increased from zero to around 60; as the median grain size became larger than 5 mm for the smooth surface grains (from the paper of ISOPE2019 [11]), the coefficient "C_{b0}" slightly decreased (shown with the dashed line). Moreover, when the rough surface grains appeared as crushed gravel, the coefficient "C_{b0}" rapidly decreased (from the new experiments for the current paper). Meanwhile, Figure 4 shows that as the uniformity coefficient became larger, the reduction coefficient "C_1" decreased the bed-load coefficient, changing from 1.0 to 0.8 and corresponding from 1 to 20 in the uniformity coefficient. Moreover, Figure 5 shows that as dry density became larger, the reduction coefficient "C_2" decreased the bed-load coefficient, changing from 1.0 to 0.25, thus corresponding from around 1.5 g/cm^3 in the dry density to 2.0 g/cm^3.

The bed-load coefficient "C_b" for Ribberink's formula can be calculated by the following equation:

$$C_b = C_{bo} \times C_1 \times C_2. \tag{10}$$

2.3. Topography Change Simulation on the Sendai-Natori Coast: A Comparison of Total Bed-Load Transport vs. Suspension Load Only

The target here was to show the impact of suspension load on topographical changes in comparison to the bed-load. For this purpose, we selected for study Japan's Sendai-Natori coast in the Miyagi Prefecture. The proposed coast is located between Sendai City in the north part and Natori City in the south part where Natori River flows west to east into the Pacific Ocean. It is noteworthy that the proposed area experienced an enormous tsunami after a 2011 earthquake with a magnitude of 9.0. The average incident tsunami height was set to 10 m on the offing boundary line.

In the numerical simulation model, the mesh size was set to 25 m and the bed-load coefficient was set to 18, as described bellows:

1. In the land area, since the median grain size was around 0.4 mm, the uniformity coefficient was around 20, and the dry density was around 1.55 g/cm^3: $C_b = C_{b0} \times C_1 \times C_2 = 22.5 \times 0.8 \times 1.0 = 18$.
2. In the beach area, since the median grain size was around 0.3 mm, the uniformity coefficient was around 10, and the dry density was around 1.55 g/cm^3: $C_b = C_{b0} \times C_1 \times C_2 = 20.0 \times 0.9 \times 1.0 = 18$.
3. In the sea area, since we could not get sufficient information, $C_b = 18$ was assumed.

The incident tsunami data on the offing boundary line of this simulation was set with reference to Figure 3.2.3 of Technical Note No.1231 of PARI (2011) [21]. However, since there was missing data for the Miyagi Prefecture tsunami, the average data between the Fukushima Prefecture and Iwate Prefecture were used. Moreover, the maximum tsunami height on the offing boundary line was estimated to be 10 m and the time of the incident phase and whole period of the first wave became 30 min and 48 min, respectively. Figure 6 shows the topography change results of the tsunami simulation considering only the impact of suspension load 48 min after calculation. The scouring depths, depicted with the pink color chart, shows a max depth of less than 0.5 m, while the deposition rate was even less than that. Ahmadi, Yamamoto, and Hayakawa (2019) [11] executed the same simulations considering the impact of both the suspension load and bed-load transport. As shown in Figure 7, the max scouring depth was found to be 4–6 m and the max deposition depth 4 m. Furthermore, to confirm the reproduction accuracy of the model, Figure 8 refers to the topographical change map from the measured data after the 2011 Great East Japan Tsunami and it is originally developed using digital elevation model (DEM). For further clarification, Figure 9 presents random spot elevation-change points along line-A on Figures 7 and 8 for comparison purposes, with the horizontal axis being the distance of the relative point from the coastline and the y-axis being the change in elevation of the point. Considering this comparison, we evaluated the topography change reproduction simulation results to be acceptable.

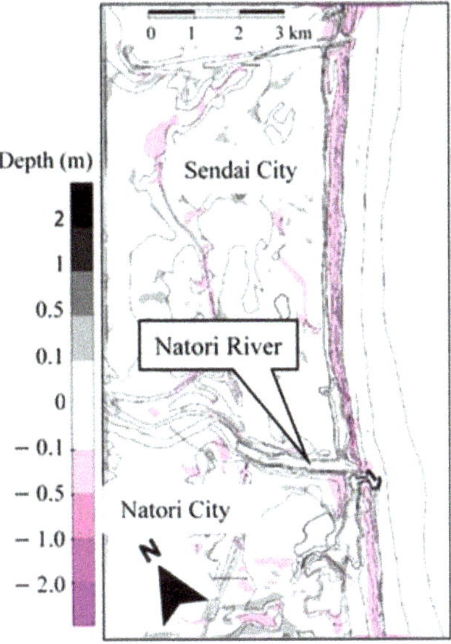

Figure 6. Topography change results (suspension load only), 48 min after calculation.

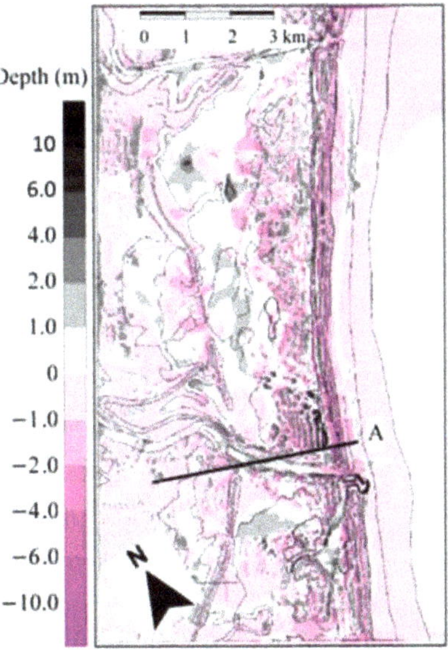

Figure 7. Topography change results (total bed-load), 48 min after tsunami calculations.

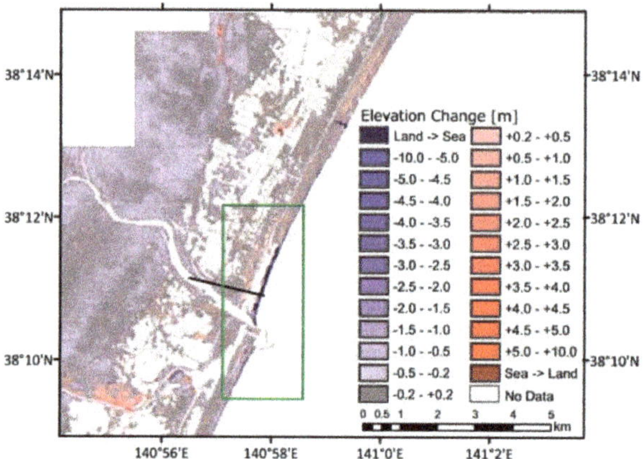

Figure 8. Topography change results from the measured data after the 2011 Tohoku tsunami, adapted from (Udo et al., 2012) [22] with permission from publisher, TAYLOR & FRANCIS, (2020).

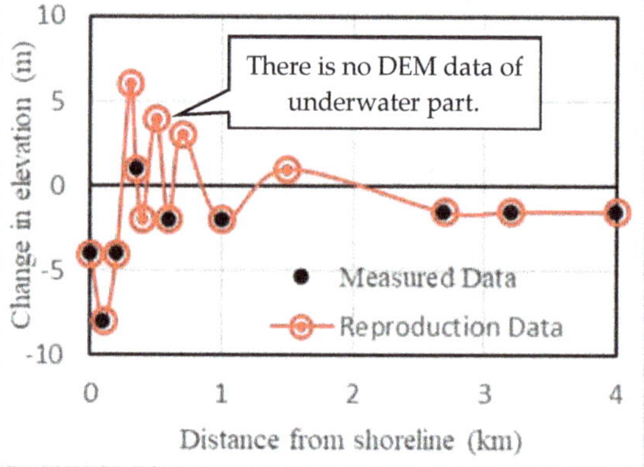

Figure 9. Relative accuracy of the topography-change along line A of the measured data vs. the reproduced simulation result.

2.4. Tsunami and Topographical Change Reproduction Simulation in Hamamatsu Coast of Japan

The Hamamatsu coast is located in the southern part of Hamamatsu City in the Shizuoka Prefecture. A tsunami simulation with conditions similar to the 2011 Great East Japan tsunami was executed on the proposed area using a numerical simulation model. Furthermore, the effects of the mitigation of the tsunami countermeasure structure development, which increased the dike height and construction of soil mounds, were also investigated. Then, the suspension load vs. total bed-load effects on topography change was presented and discussed.

The tide level data for the 2011 Great East Japan tsunami, recorded by the Japanese government, is shown in Figure 10a. Because the proposed coast for our simulation (Figures 6–9) was the Miyagi Prefecture's northern part, we used the fourth data (orange curve) from the top of Figure 10a. However, since there was no tidal data after a peak at Miyagi Prefecture's northern part, we supplied a portion of tidal data after the peak with average values using the data of the Iwate southern part (red curve) and the Fukushima Prefecture part (green curve). At the time, in order to eliminate reflective waves and harassing waves, we eliminated waves where cycles were smaller than the cycle of the original tsunami waves from the epicenter. The results are shown in Figure 10b.

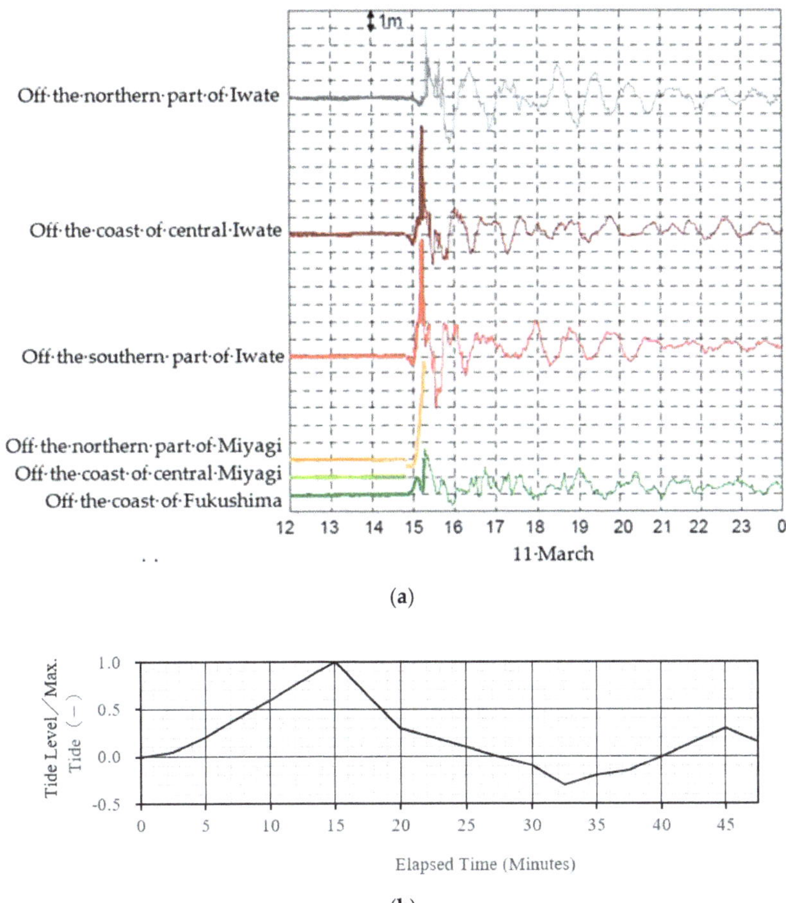

(a)

(b)

Figure 10. (a) Tide level data of the 2011 Great East Japan tsunami (from No.1231 Technical Report of the port and airport research institute of the Japanese Government); (b) input data of off-shore tide levels observed off of the Miyagi Prefecture. Adapted from Figure 3.2.3 of Technical Note No.1231 of PARI 2011) [21] with permission from the publisher, Japan National Research and Development Institute, (2020).

By using the tide waveform shown in Figure 10b and inputting 8 m as the maximum tide from the offshore boundary (water depth 100–110 m), the maximum tsunami heights on the shoreline became 10 m–15 m.

The bed-load coefficient was set to 18 because the sediment characteristics were similar to that of the Sendai-Natori coast. Figure 11 shows the topography of the proposed area and the calculation range. As depicted, the Tenryu River located on the east side of Hamamatsu City flows north to south toward the Pacific Ocean. Figure 12 shows the building ratio in a calculation mesh. The yellow color (the building ratio is 1% in a mesh area) refers to the soil and sand area, whereas the green (5% in a mesh area) refers to the wood and forest area. Further, the light brown (30% in a mesh area) shows a house with a garden, whereas the brown area (70% in a mesh area) shows a building area. The calculation mesh size was set to 12 m.

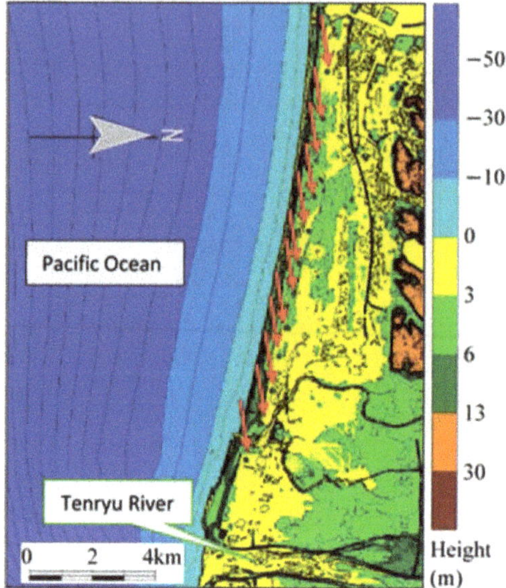

Figure 11. Topography map and calculation range of the Hamamatsu Coast.

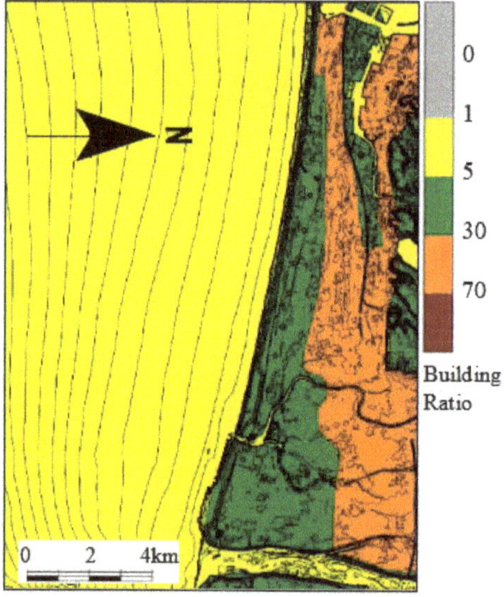

Figure 12. Building ratio in a calculation mesh (the area ratio of buildings and trees in a mesh).

According to the simulation results, 16 min after beginning calculations, the tsunami height on the shoreline reached its maximum levels for a wide range. Afterward, for about 10 min, the maximum inundation depth was 5 m in a wide range in the inland area.

Figures 13–15 show the inundation depth, tsunami velocity, and topographical change results, respectively, 33 min after the start of calculations when the seawater intrusion mode was over. According to these figures, the inundation area extended 3–6 km from the coastline. The forward velocity was calm after 33 min from the start of calculations and the maximum scouring depth was about 6 m.

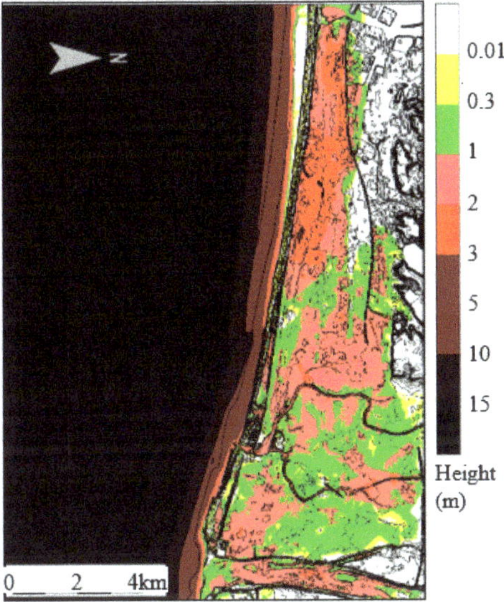

Figure 13. Inundation depths 33 min after the start of calculation (present condition).

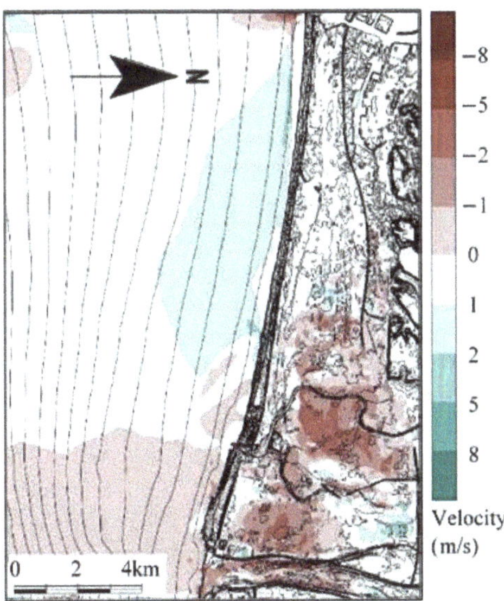

Figure 14. Tsunami velocity 33 min after the start of calculation (present condition).

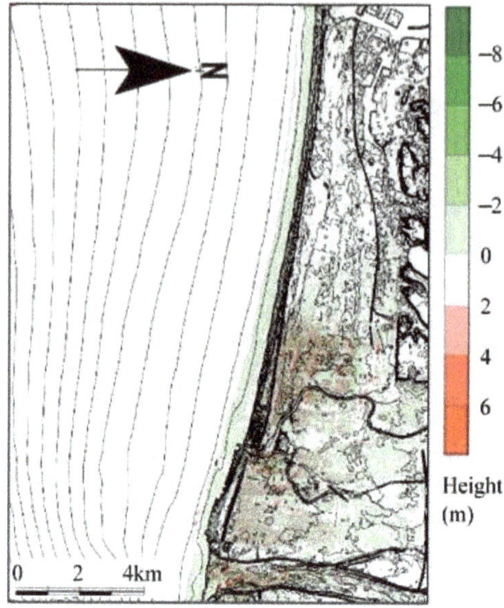

Figure 15. Topography change output 33 min after the start of calculation (present condition).

In the proposed area, as countermeasures to a great tsunami, the crest height of the coastal embankment increased from TP +7 m to TP +13 m after the construction of CSG embankments and evacuation soil mounds; their locations are pointed out with red arrows in Figure 11. Here, including these developments in the simulation criteria, the output results significantly changed. The new CSG embankment did not wash out due to the increased crest height. The maximum inundation depth was halved over a wide range (about 2.5 m). Figures 16–18 show the calculation results 33 min after the start, demonstrating that the inundation width from the coastline decreased to about half (1.5–3 km from the shoreline). The maximum inundation depth also decreased to about 2 m except around the Tenryu River and Magome River, where there was no CSG embankment. The tsunami velocity decreased by half later in the calculation time. Furthermore, the 12 soil mounds constructed with a circular shape of 100 m diameter (top) and 150 m diameter (bases) with a height of 8 m were expected to function as evacuees because the scouring depth around them was less than 1 m in depth. However, since they were made of clay and tended to weather easily and lose its shape, plantation cover was thought to potentially overcome this issue.

The next objective was to compare the suspension load and total sediment transport load topography changes. Figures 19 and 20 show that the maximum scouring depth 28 and 33 min from the start of calculations must be around 0.5 m to 1 m, respectively, which was less than that of the total bed-load transport (i.e., 4–6 m in depth, as shown in Figure 17). It is noteworthy that the simulation input criteria for this comparison were similar as the ones defined above.

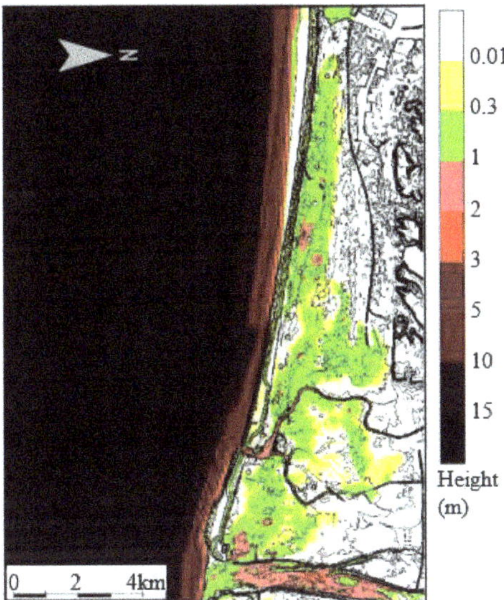

Figure 16. Inundation depths 33 min after the start of calculation (i.e., after structure developments).

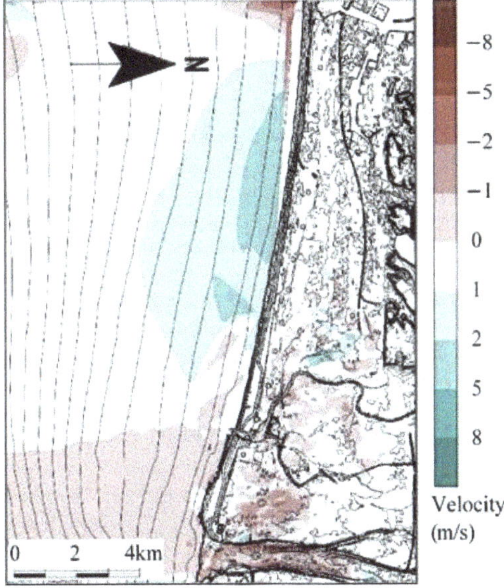

Figure 17. Tsunami velocity 33 min after the start of calculation (after structure developments).

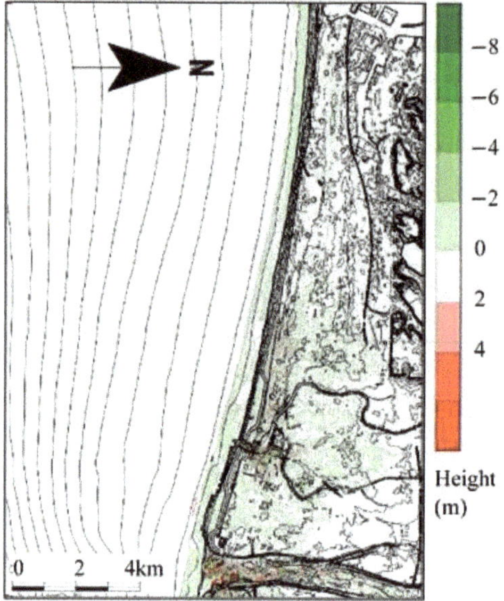

Figure 18. Topography change output 33 min after the start of calculation (after structural developments).

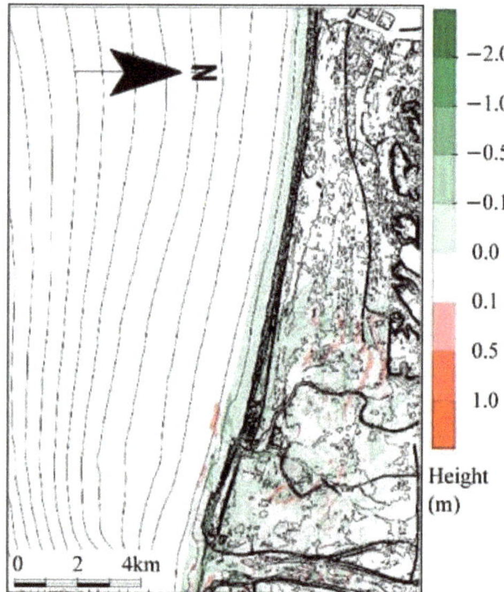

Figure 19. Topography change due to suspension load only, 28 min after the start of calculation (present condition).

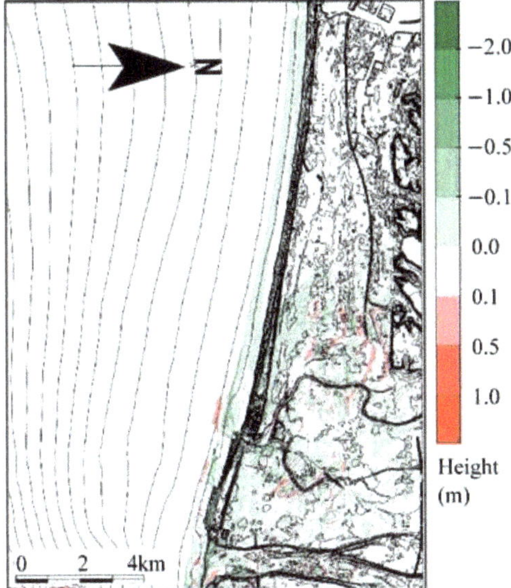

Figure 20. Topography change due to suspension load only, 33 min after the start of calculation (present condition).

3. Model of Building Destruction Phenomenon by a Huge Tsunami

Many researchers have already published fragility curves for structures due to tsunami loads. However, since their methods are mostly based on statistics and probability models—such as fragility curves by Nanayakkara and Dias (2016) [23], as well as Suppasri et al. (2012, 2013, 2015) [24–26]—their fragility curves can be used to easily evaluate destruction probabilities of various building groups. Further, they are strongly influenced by the characteristics of the collected data.

On the other hand, the fragility curves presented in our paper can be used for a stability check of each building by using its structural parameters, such as pillar width, pillar interval, number of stories, and so on. Moreover, since the fragility curves in our paper were made via stress calculations to main members, our method can consider the influence of the strength for the materials of various main members.

3.1. Field Survey of Building Destruction by the 2011 Great East Japan Tsunami

In April 2011, our team executed an investigation of buildings destroyed by a huge tsunami caused by an earthquake off the Pacific coast of Tohoku where wooden buildings suffered heavy damages when inundation depths exceeded 1.5 m and reinforced concrete buildings broke when inundation depths exceeded 5 m. Using the investigation data presented later in this paper, our team developed diagrams to predict a building's destruction grade and compared it against the real case site data. Therefore, the proposed diagrams explained the actual building damage. Moreover, the inundation depths taken from the numerical simulation model were used alongside the proposed diagrams to predict the actual building destruction with sufficient accuracy.

The Great East Japan Earthquake tsunami on March 11, 2011 caused enormous inundation damage along the Pacific coast from the Aomori to Chiba prefectures, and many damage investigation results have been already reported. In particular, the Institute of Industrial Science at the University of Tokyo (2011) studied details of the tsunami's force evaluation method, the destructive tsunami pressure of reinforced concrete buildings and concrete structures, the impact force of drifting objects, and so on, to review the evacuation building design method. However, there was no universal failure limit investigation report to consider the differences in dimensions of a structure's main components. Therefore, because the Nankai tsunami is expected to occur in the near future, and further research on disaster prevention and mitigation is desired, the authors focused on structural damage caused by the 2011 tsunami. Investigations and comparisons with the compiled structural damage data show that the structural damage limit calculation diagrams of Yamamoto et al. (2011) [13] and two-dimensional run-up numerical simulation model can be used to predict structural damage grade.

From 9–15 April 2011, a building destruction investigation including the destruction status of reinforced concrete buildings, wooden houses, concrete block walls, and steel-framed buildings due to the tsunami inundation was executed. The investigation was accomplished in the southern part of Miyako City, Iwate Prefecture, and the northern part of Hitachi City, Ibaraki Prefecture (excluding the area around Ishinomaki and the off-limits area due to the Fukushima Daiichi nuclear accident). Tables 2–4 show the details of the investigations including the latitude, longitude, inundation depth, column and wall materials, basic dimensions, and destruction status from the actual devastated area. Figures 21–27 show the typical example photographs of heavy destruction taken during the aforementioned surveying.

Figure 21. Reinforced concrete house, inundation depth 5.5 m, Yamada-cho.

Figure 22. Reinforced concrete building, inundation depth 10 m, Rikuzentakata City.

Figure 23. Steel building, inundation depth 7.5 m, panel wall destruction, Yamada-cho.

Figure 24. Reinforced concrete building, inundation depth 8.5 m, Rikuzentakata City.

Figure 25. Reinforced concrete building, inundation depth 6.5 m, Ofunato City.

Figure 26. Wooden house, inundation depth 2 m, wall destruction, Kitaibaraki City.

Figure 27. Block wall, inundation depth 2 m, wall fail, Kitaibaraki City.

Table 2. The damage survey for reinforced concrete buildings during the 2011 Great East Japan Earthquake tsunami.

Prefecture	Iwate Prefecture									
City Name	Miyako City				Yamada-Cho			Kamaishi City		
Latitude	39°38′28″	39°38′25″	39°38′22″	39°28′17″	39°28′17″	39°28′17″	39°28′33″	39°28′33″	39°18′37″	39°15′59″
Longitude	141°52′46″	141°57′56″	141°58′10″	141°57′15″	141°57′14″	141°57′14″	141°57′17″	141°57′17″	141°53′19″	141°53′11″
Inundation depth (m)	4.8	5.0	6.0	6.0	5.5	5.5	6.5	6.5	5.0	6.5
Floor number	2	2	2	3–4	3	2	4	2	2	4
Column width (m)	0.44	0.63	0.85	0.50	0.55			0.49	0.53	0.65
Column height (m)	3.3	3.2	3.5	3.2	4.5			4.0	3.5	3.5
Column spacing (m)	6.0	6.0	5.5	6.5	6.0			4.0	4.5	5.0
Column break	None	None	None	None	None			None	None	None
Wall thickness (m)					0.3	0.25	0.24	0.10	0.17	0.30
Wall height (m)					4.0			3.5	3.2	
Wall break	None	None	None	None	None	None	None	Broken	Broken	None

Prefecture	Iwate Prefecture									
City name	Ofunato City						Rikuzentakata City			
Latitude	39°4′1″	39°4′1″	39°4′1″	39°3′58″	39°4′7″	39°0′36″	39°0′32″	39°0′52″	39°0′51″	39°0′51″
Longitude	141°43′18″	141°43′18″	141°43′14″	141°43′15″	141°43′19″	141°38′38″	141°38′42″	141°38′22″	141°38′02″	141°38′02″
Inundation depth (m)	6.5	6.5	5.0	6.5	8.0	10.0	9.0	8.5	8.5	8.5
Floor number	3	2	2	2	3 – 4	5	2	3	2	2
Column width (m)	0.65	0.45	0.75	0.40	0.75		0.75	0.61	0.60	0.55
Column height (m)	3.5	3.1	3.4	3.2	4.0		3.5	3.5	3.7	3.5
Column spacing (m)	4.8	5.2	5.8	3.5	15.0		6.6	9.4	3.8	5.0
Column break	None	None	None	None	None		None	None	None	None
Wall thickness (m)	0.23	0.21	0.20	0.20	0.20	0.27	0.20	0.17	0.18	0.22
Wall height (m)	3.4								3.5	

Table 2. *Cont.*

Wall break	Partial destruction	None	None	None	None	None	None	None	Broken	None
Prefecture	Miyagi Prefecture						Fukushima Prefecture	Ibaraki Prefecture		
City name	Kesennuma City				Sendai City		Soma City	Kitaibaraki City		Hitachi City
Latitude	38°54′57″	38°54′58″	38°54′44″	38°16′28″	38°13′23″	38°13′20″	37°49′38″	36°49′49″	36°47′44″	36°34′41″
Longitude	141°34′55″	141°34′58″	141°34′51″	141°0′20″	140°59′05″	140°58′51″	140°58′23″	140°47′32″	140°45′15″	140°39′28″
Inundation depth (m)	8.0	7.0	7.2	5.0	7.5	7.5	6.0	2.7	2.0	3.5
Floor number	3	2	2	2	2	4	2	2 – 3	2	2
Column width (m)	0.40	0.76	0.52	0.82	0.77	0.60	1.20	0.44	0.27	0.90
Column height (m)	2.7	4.0	3.5	3.7	3.0	3.4	3.5	3.4	3.1	4.0
Column spacing (m)	3.5	5.0	3.8	6.5	7.4	4.5	20.0	3.6	3.7	6.0
Column break	None	None	None	None	None	None	None	None	None	None
Wall thickness (m)	No wall				0.21	0.20	Without Walls	0.20		0.23
Wall height (m)								3.2		3.5
Wall break		None	None	None	None	None		None	None	None
Note	The column spacing is a value parallel to the tsunami penetration direction.									

Table 3. Wooden house damage survey results for the 2011 Great East Japan Earthquake tsunami.

Prefecture	Iwate Prefecture				Miyagi Prefecture		Fukushima Prefecture	Ibaraki Prefecture			
City name	Miyako City				Kesen-numa	Sendai City	Iwaki City	Kitaibaraki City			Hitachi City
Latitude	39°38′28″	39°38′28″	39°38′28″	39°38′28″	38°54′18″	38°14′11″	36°51′39″	36°49′49″	36°47′44″	36°47′29″	36°34′41″
Longitude	141°52′45″	141°52′45″	141°52′45″	141°52′46″	141°34′58″	140°57′39″	140°47′25″	140°47′32″	140°45′15″	140°44′59″	140°39′27″
Inundation depth (m)	2.9	2.2	2.9	3.7	2.7	0.95	2.0	2.0	2.0	1.3	1.2
Floor number	2	2	2	2	2	2	2	2	1	2	2
Pillar width (m)	0.34	0.12	0.12	0.15	0.12	0.14	0.20	0.20	0.15	0.12	0.12
Pillar height (m)	3.0	2.7	3.0	3.0	2.7	2.7	2.7	3.1	2.7	2.7	2.7
Pillar interval (m)	6.2	1.8	4.8	2.5	1.8	1.9	3.6	3.6	3.7	3.7	3.6
Pillar break	None	None	Broken	Broken	Broken	None	None	None	None	None	None
Wall structure	Board + metal		Board + metal				Board + metal	Earth + bamboo + board	Earth + bamboo + board		Board + metal
Wall thickness (m)	0.15		0.12				0.10	0.15	0.10		0.10
Wall height (m)	2.9		2.9				2.6	3.0	2.5		2.6
Wall break	Broken	Broken	Broken	Broken	Broken	None	Broken	Partial destruction	Broken	None	Partial destruction
Note	The column spacing is parallel to the tsunami penetration direction.										

Table 4. Survey on damage to concrete block walls during the 2011 Great East Japan Earthquake tsunami.

City Name	Miyako City		Sendai City			Kitaibaraki City			
Latitude	39°36′54″	38°16′41″	38°16′41″	38°14′11″	38°13′46″	36°49′49″	36°47′44″	36°47′29″	36°47′29″
Longitude	141°57′27″	140°59′23″	140°59′23″	140°57′39″	140°58′19″	140°47′32″	140°45′15″	140°44′59″	140°44′59″
Inundation depth (m)	0.5	1.0	1.0	0.9	2.0	2.0	2.0	1.3	1.3
Wall structure	Black Mass Concrete	Perforated rebar Concrete block	Black Mass Concrete	Reinforced concrete Block	Reinforced concrete Block	Reinforced concrete Block	Reinforced concrete Block	Reinforced concrete Block	Black Mass Concrete
Break of the wall	None	None	Broken	None	Broken	Broken	Broken	Broken	Broken

Considering the above tables and photographs, we considered the breakdown characteristics of various buildings after the tsunami. The results are as follows:

(a) In the case of a reinforced concrete building, windows and doors were torn when immersed in water at more than half of their surface area, but the walls only broke when the window occupancy ratio was 30% or less, wall thickness was 23 cm or less, and inundation depth was 5 m or more. Although there were no column failure cases for old buildings with insufficient seismic design, as is the case for Onagawa-Cho, the foundation was broken and the entire building was overturned.

(b) In the case of a wooden house, when the inundation depth became about 1.2 m or more, windows and doors were easily damaged, and the 10 cm thick walls started to break. Pillars were more likely to collapse when the inundation depth exceeded 2.5 m.

(c) Concrete block walls (standard thickness 10 cm) were more likely to fall if the inundation depth was 1 m or more when there was no reinforcement and 1.3 m or more when there was a reinforcing bar.

(d) For steel-framed buildings, the data could not be listed due to space limitations, but there were no cases where the main steel frame was broken except for the case where large drifting objects such as ships and automobiles collided with them. However, since the wall was panel-shaped and thin, the wall would likely break when inundation depth was about 3 m or more.

Evaluating tsunami-related building destruction was performed with high accuracy using the conventional building design method, but the cost was high (US$1000 or more per house). The cost was due to the need for experts to conduct such complicated examinations. It is extremely difficult to formulate a disaster prevention and mitigation plan assuming damage from the design method. On the other hand, there is a simple method for examining the relationship between inundation depth (and inundation flow velocity) and the degree of destruction for each type of structure based on actual damage data. The effect on the degree is not known. Therefore, Yamamoto et al. (2011) [13] proposed a simple evaluation method, which is described below.

A typical building is designed to support loads with a rigid frame structure consisting of two columns and one beam. Furthermore, according to a hearing survey on the destruction of buildings caused by the 1993 Hokkaido Nansei-Oki Earthquake tsunami and the 2004 Indian Ocean Tsunami, when a tsunami large enough to destroy buildings occurs, windows and doors are instantaneously damaged, and water damages the inside and outside of buildings. When the static water pressure is offset, walls act openly without a break, and when the column is destroyed, it breaks from the base and becomes "destroyed". If the column is not destroyed, it is in a "half-broken" state. Since the destruction was the same during the 2011 tsunami, the pillars were integral with the base and were stronger than the walls. Under the condition that walls have windows and doors that break instantaneously, we assumed the presence of a gate-type rigid frame where the tsunami force entered through the top load on the roof, as shown in Figure 28. Thus, the relationship between the inundation depth and unbreakable critical column width can be obtained from the stress at the column base where the tsunami force acts directly.

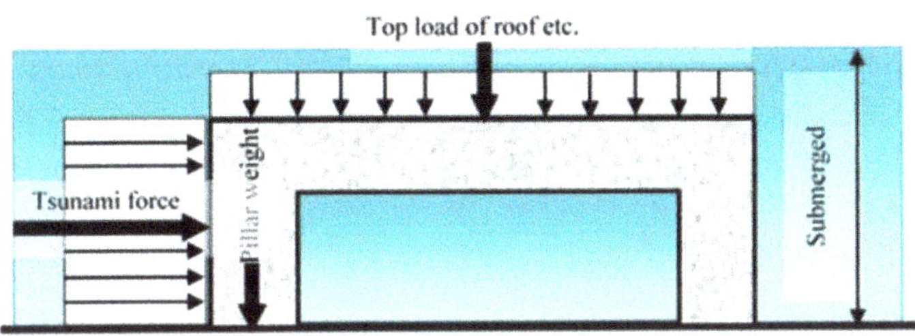

Figure 28. Illustration of the Ramen Gate Model.

Tsunami force calculation equations include Asakura et al.'s (2000) [27], which was used near the coastline with a large Froude number, as well as Iizuka and Matsutomi's (2000) [28] formula in the inland area with reduced momentum. The Institute of Industrial Science at the University of Tokyo (2011) proposed a considerably safe tsunami force calculation standard for evacuation building design. Furthermore, since the dimensions of a building's main members are standardized by the representative dimensions of humans, combinations of basic dimensions such as column spacing and column height are limited. This chapter introduces threshold diagrams for calculating the column width and wall thickness, which break the limits for a given inundation height and a limited combination of frequently used for column spacing and column height.

As shown in Figure 28, we used a gate-type Rahmen model (i.e., a height of three meters and an interval length between two pillars of five meters as typical values) for the building structure of the on-off shore direction and assumed that the incident tsunami pressure acting on the building's seaside wall could be obtained using Equation (9) (Iizuka and Matsutomi (2000)) [28] and that stress acted on the base of the seaside pillar.

$$F = \frac{C_f}{2} \rho B h v^2 = \frac{C_f}{2} \rho B h \left(F_r \sqrt{gh}\right)^2 \tag{11}$$

where F is the tsunami force, C_f is the hydrodynamic coefficient (= 2), ρ is the seawater density, B is the tsunami force action width, h is the inundation depth in front of the building, v is the inundation velocity, F_r is the Froude number (= 1.1), and g is the gravitational acceleration.

3.2. Confirmation of the Proposed Conventional Method with Disaster Data

3.2.1. Threshold Width of Columns in Reinforced Concrete Buildings

From the frequency characteristics in the actual disaster data, the column spacing of the first floor ramen part is an integral multiple of half (≒ 1 yard) + rounded value. The column height of the first-floor ramen section was 3.35 m and the median value of the high frequency and cross-sectional structure of the columns was a rectangular cross-section of double reinforcement bars in which reinforcing rods were arranged on the seaside and landside because of the high frequency. In the case of two and five stories, the bending/shear stress at the base of the column was calculated by changing the tsunami's inundation depth and by comparing it with member strength. Thus, we determined the relationship between the inundation depth and the critical width that would not break, as shown in Figure 29. Here, to obtain a result on the safe side, the upper part of the second floor was considered as the upper load and was modeled with a slightly excessive concrete mass of 6.4 m × column spacing × (0.25 m × number of floors). The reinforcement ratio (= the steel ratio) was 0.05 close to the lower limit of the ratios and the clearance thickness (= the cover thickness) was 5.0 cm from the outer concrete surface so that it minimized the effective column width in the collected data. The effect of a band rebar (= a lateral tie, a tie hoop) was ignored. The compressive bending strength of concrete was considered 20 N/mm² , which was close to the lower limit. The shear strength was considered 1/10 of that and the tensile bending strength of reinforcing bars was considered 300 N/mm², which was also close to the lower limit.

The non-destructive data of the two- and three-story buildings in Table 2 are plotted in Figure 29a, and the four- and five-story buildings show in Table 2 are plotted in Figure 29b with white circles (unit: meter). The applicability of this calculation diagram was high because the curve corresponded to the column spacing of each white circle and was approximately at the lower right. Under this condition, the type of failure was mainly the tensile bending failure of the reinforcing bar. In the case of seismic design of six stories or more, the possibility that the tsunami force exceeded the seismic force was low, and the necessity of the tsunami stability study became low as well.

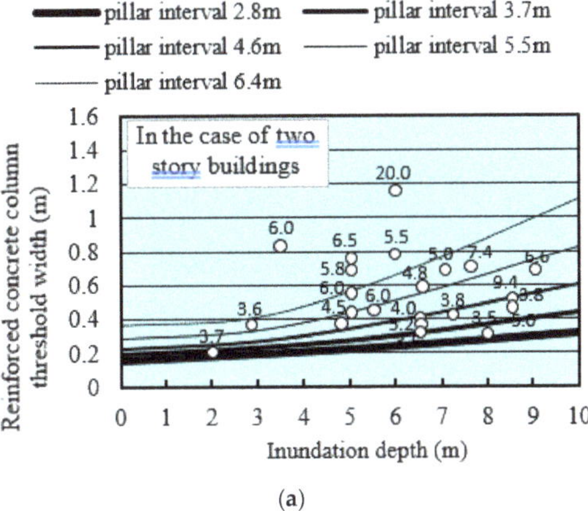

(a)

Figure 29. *Cont.*

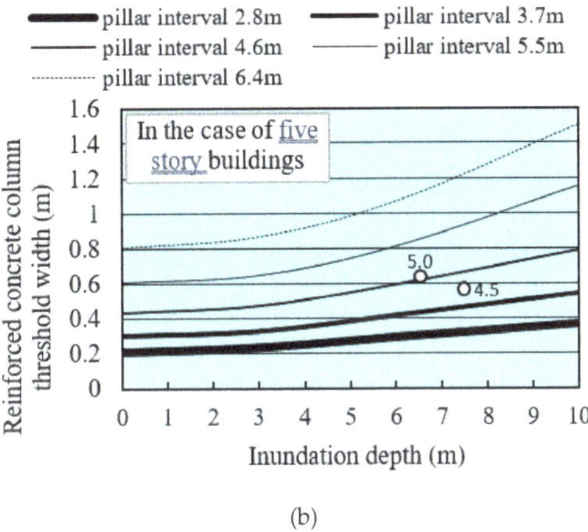

(b)

Figure 29. (**a**) Relationship between inundation depth and critical width of reinforced concrete columns (the numbers in the figure are the pillar intervals). (**b**) Relationship between inundation depth and critical width of reinforced concrete columns (the numbers in the figure are the pillar intervals).

3.2.2. Threshold Width of the Pillar in Wooden Buildings

Yamamoto et al. (2011) [13] showed the applicability of the gate-type ramen for the wooden pillars of wooden houses, using the data from the 1993 Hokkaido Nansei-Oki earthquake tsunami, as shown in Figure 30a,b. The height of the column on the first floor was set at 3.3 m, which was higher for safety, and the cross-section of the column was rectangular. The top load (i.e., the roof weight) of a wooden building consisted of the pillar interval × the pillar interval × 0.1 m of a wood portion (density 500 kg/m^3) and the pillar interval × the pillar interval × 3.5 cm of a roof tile portion (density of 2000 kg/m^3). In the case of a two-story building, a portion other than the roof on the second floor was modeled as a wooden lump with the pillar interval × the pillar interval × 0.4m. Regarding the bending strength and shear strength of wooden pillars, 20 N/mm^2 and 2.4 N/mm^2 of intermediate materials were used, respectively, referring to Notification 1452 of the Ministry of Construction (2000) [29].

Figure 30a,b also plots the damage data shown in Table 2, adding the pillar interval (unit: m). The open circles are non-destructive, the gray circles are partial destruction cases, the black circles are all destruction cases, and the curves corresponding to the pillar interval of each white circle are on the lower right, whereas the curves for each gray and black circle are on the upper left. It can be said that the strength of the members was appropriately evaluated.

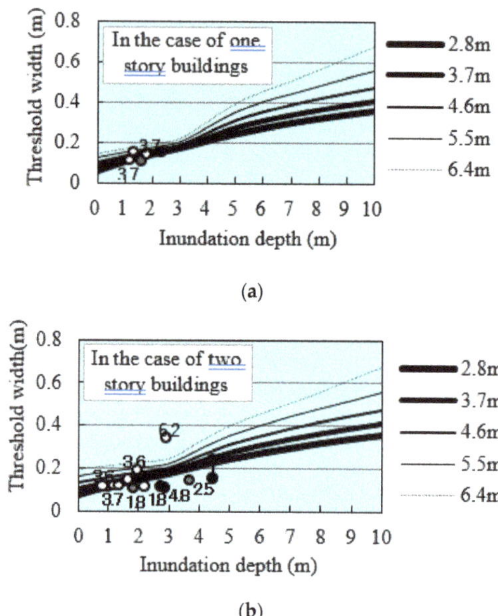

(a)

(b)

Figure 30. (a) Relationship between inundation depth and critical width of wooden pillars (the numbers in the figure are the pillar intervals). (b) Relationship between inundation depth and critical width of wooden pillars (the numbers in the figure are the pillar intervals).

3.2.3. Threshold Inundation Depth for Concrete Block Walls

For a wall stacked with blocks (standard thickness 10 cm × length 40 cm), the inundation depth of the unbreakable limit for each concrete strength were examined when only the tsunami force acted on the cantilever. In that case, the seaside horizontal hydrostatic pressure could not be ignored and the results are shown Figure 31a,b. Figure 31a shows the case where there was no reinforcement and Figure 31b shows the case where one reinforcing bar with a diameter of 1.3 cm was inserted for each block. Under these conditions, the wall broke by the tensile bending stress of the reinforcing bar and a value 15 times the compressive bending strength of concrete was used for this strength. These figures show the damage data of the Hokkaido Nansei-Oki earthquake tsunami. The open circles are non-destructed, the closed circles are all destructed cases, and the open circles are below the curve, whereas the black circles are above, indicating that the applicability of this calculation diagram is high.

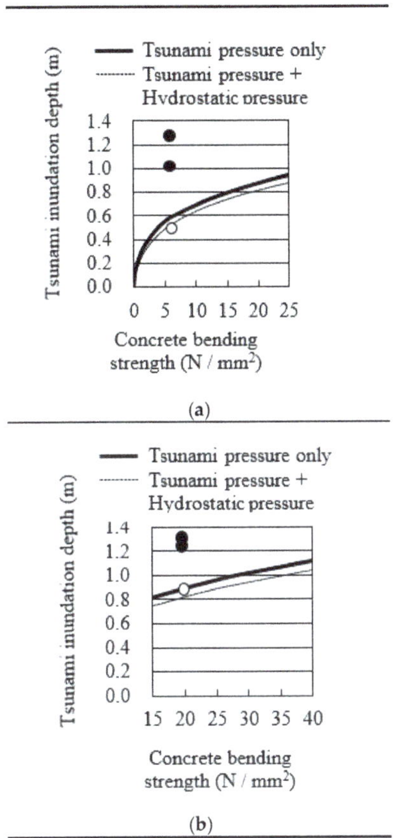

Figure 31. (**a**) Tsunami inundation depth of the limit that does not break (in the case without steel rods). (**b**) Tsunami inundation depth of the limit that does not break (in the case with steel rods).

3.3. Verification Examples of Building Destruction Due to 2011 Great East-Japan Tsunami

Using the tsunami simulation model and the developed threshold diagrams for building destruction design, some verification examples are presented in Table 5. Figures 32–34 show the calculated maximum inundation depths in the three proposed areas: Rikuzentakata City, Kesennuma City, and Sendai City. Referring to verification example results, the design destruction state matched the real destruction state of buildings except for the first one. Although the reinforced concrete building located in Rikuzentakata City was stated as "broken" in the design, it was not broken in the real situation and was still on the design's safe side. Furthermore, when comparing the calculated inundation water depths and measured depths, we observed that the results were acceptable.

Table 5. Verification examples of building destruction due to the 2011 Great East Japan tsunami.

City Name	Building Type	Calculated Water Depth (m)	Structure Properties			Design Destruction State	Measured Water Depth (m)	Real Destruction State
			Number of Stories	Column Width (m)	Column Spacing (m)			
Rikuzentakata	RC * Museum	8.80	2	0.75	6.6	Broken	9.0	Not broken
Kesennuma	RC * House	7.50	2	0.52	3.8	Not broken	7.2	Not broken
Kesennuma	RC * House	7.80	3	0.4	3.5	Not broken	8.0	Not broken
Kesennuma	RC * House	7.30	2	0.76	5.0	Not broken	7.0	Not broken
Kesennuma	Wooden House	2.60	2	0.12	1.8	Broken	2.7	Broken
Sendai	Wooden House	1.00	2	0.14	1.9	Not broken	0.95	Not broken
Sendai	R.C * Block Wall	2.50				Broken	2.0	Broken
Sendai	R.C * Block Wall	0.90				Not broken	0.9	Not broken

* RC: reinforced concrete.

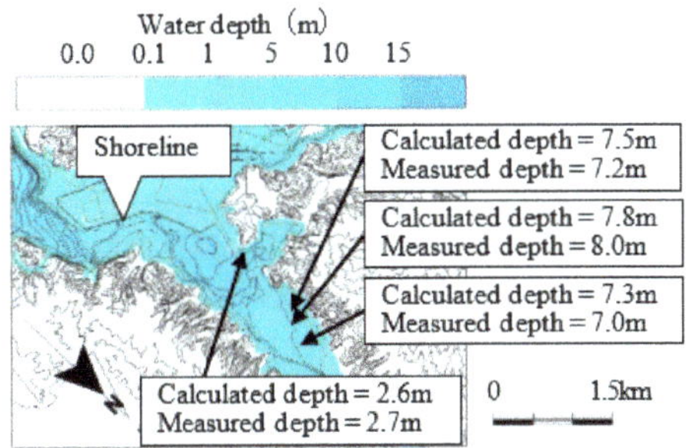

Figure 32. Flood situation in Kesennuma Coast 25 min after arrival of the 2011 huge tsunami.

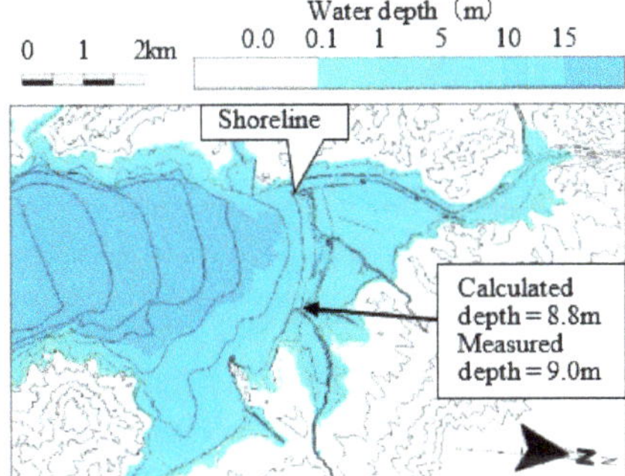

Figure 33. Flood situation in Rikuzentakata Coast 31 min after arrival of the 2011 huge tsunami.

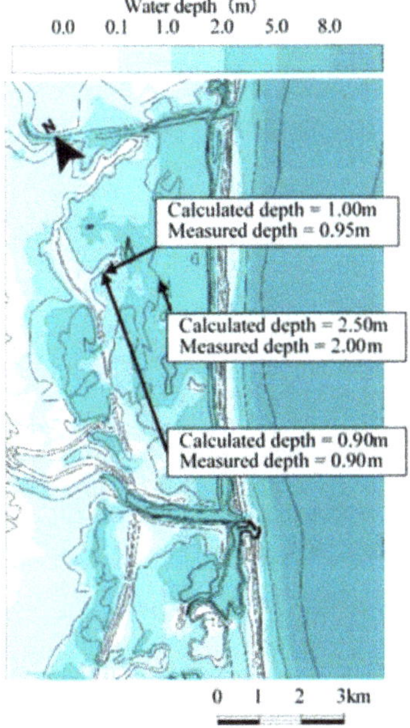

Figure 34. Flood situation in Sendai Coast 48 min after arrival of the 2011 huge tsunami.

4. Conclusions

To acquire the bed-load coefficient of Ribberink's bed-load transport equation, we developed useful diagrams. These diagrams, illustrated in Figures 3–5, present a wide range of grain size, dry density, and uniformity coefficients. Figure 3 shows that the median grain size (i.e., natural soil, sand, and gravel) increased to around 5 mm and the bed-load coefficient increased from zero to around 60. Further, when the median grain size became larger than 5 mm for smooth surface grains, the bed-load coefficient remained almost constant; however, for the rough surface grains, such as crusher-run stone, the bed-load coefficient decreased to around 5 mm. Meanwhile, Figure 4 shows that as the uniformity coefficient became larger, the reduction coefficient for decreasing the bed-load coefficient changed from 1.0 to 0.8, which corresponded from 1 to 20 in the uniformity coefficient. Moreover, Figure 5 shows that as dry density became larger, the reduction coefficient for decreasing the bed-load coefficient changed from 1.0 to 0.25, which corresponded with 1.5 g/cm^3 in the dry density to 2.0 g/cm^3.

From the Sendai-Natori coast tsunami reproduction simulation results, we observed that the impact of suspension load versus total bed-load was not large. The results of suspension only showed a max depth of scouring less than 0.5 m while that of the total bed-load was around 4–6 m. Furthermore, when we researched to the deposition depths, the deposition depth with suspension loads was negligible, especially when compared to the total bed-load, which was around 5–6 m.

The results of the tsunami simulation on the Hamamatsu coast, where evacuation soil mounds were constructed and the crest height of the coastal dikes increased from TP +6 m to TP +13 m, showed that the evacuation soil mounds served their purpose; however, because they were directly affected by weathering, plantations could overcome this issue. Moreover, the dike improvements significantly decreased inundation depth and height, while inundation water intruded into the area around Magpome River and Tenryu River. Still, however, there were no riverbank improvements.

Lastly, using the building destruction survey data from the 2011 Great East Japan tsunami, useful diagrams were developed and presented to significantly decrease the cost and time needed to evaluate building destruction analysis due to a tsunami. These diagrams are shown in Figures 29–31 and confirmed that the tsunami damage prediction method based on the method of calculating the threshold limit of columns and walls was useful.

Author Contributions: Conceptualization, S.M.A., Y.Y. and V.T.C.; methodology, S.M.A. and Y.Y.; software, S.M.A. and Y.Y.; validation, S.M.A., Y.Y. and V.T.C.; formal analysis, S.M.A. and Y.Y.; investigation, Y.Y. and S.M.A; resources, Y.Y.; data curation, V.T.C.; writing—original draft preparation, S.M.A.; writing—review and editing, S.M.A; visualization, supervision, Y.Y.; project administration, Y.Y.; funding acquisition, Y.Y. All authors have read and agreed to the published version of the manuscript.

Funding: This research was executed under the support of Grants-in-Aid for Scientific Research (c; 18k04667) of JSPS. We express our deepest gratitude.

Conflicts of Interest: The authors declare no conflict of interest.

References

1. Takahashi, T.; Imamura, F.; Shuto, N. Numerical Simulation of Topography Change Due to Tsunamis. In Proceedings of the IUGG/IOC International Tsunami Symposium, Wakayama, Japan, 23–27 August 1993; pp. 243–255.
2. Takahashi, T.; Imamura, F.; Shuto, N. Examination of the applicability and reproducibility of a tsunami move model. In Proceedings of the Coastal Engineering, Venice, Italy, 4–9 October 1993; Volume 40, pp. 171–175. (In Japanese).
3. Kobayashi, A.; Oda, Y.; Higashie, T.; Takao, M.; Fujii, N. Study on Sand Movement due to Tsunami. In Proceedings of the Coastal Engineering, JSCE, Orlando, FL, USA, 2–6 September 1996; Volume 43, pp. 691–695. (In Japanese).
4. Fujii, N.; Ohmori, M.; Takao, M.; Kanayama, S.; Ohtani, H. On the Deformation of the Sea Bottom Topography due to Tsunami. In Proceedings of the Coastal Engineering, JSCE, Copenhagen, Denmark, 22–26 June 1998; Volume 45, pp. 376–380. (In Japanese).

5. Takahashi, T.; Shuto, N.; Imamura, F.; Asai, D. Modeling Sediment Transport due to Tsunamis with Exchange Rate between Bed Load and Suspended Load Layers. In Proceedings of the 27th International Conference on Coastal Engineering (ICCE), Sydney, Australia, 16–21 July 2000; pp. 1508–1519.

6. Nishihata, T.; Tajima, Y.; Moriya, Y.; Sekimoto, T. Topography Change due to the Dec 2004 Indian Ocean Tsunami. In Proceedings of the International Conference on Coastal Engineering (ICCE), San Diego, CA, USA, 3–8 September 2006; pp. 1456–1468.

7. Kihara, N.; Matsuyama, M. Hydrostatic Three Dimensional Numerical Simulations of Tsunami Scour in Harbor. *Annu. J. Coast. Eng. JSCE* **2007**, *54*, 516–520. (In Japanese) [CrossRef]

8. Nakamura, T.; Mizutani, N. Simulation on Tsunamiinduced Topographic Change Using Sediment Transport Formula with Dynamic Effective Stress inside Sand Bed. *Annu. J. Civ. Engg. Ocean JSCE* **2008**, *24*, 57–62. (In Japanese)

9. Ca, V.T.; Yamamoto, Y.; Charusrojthanadech, N. Improvement of Prediction Methods of Coastal Scour and Erosion due to Tsunami Back-flow. In Proceedings of the 20th International Offshore and Polar Engineering Conference, Beijing, China, 20–25 June 2010; pp. 1053–1060.

10. Ahmadi, S.M.; Yamamoto, Y.; Miyake, R. Examination to Some Problems on the Prevention and Mitigation of Tsunami Disaster. In Proceedings of the 28th International Offshore and Polar Engineering Conference, Sapporo, Japan, 10–15 June 2018; pp. 701–708.

11. Ahmadi, S.M.; Yamamoto, Y.; Hayakawa, M. Development of a Rational Prediction Method of Topographical Change by a Tsunami. In Proceedings of the 29th International Offshore and Polar Engineering Conference, Honolulu, HI, USA, 16–21 June 2019; pp. 3226–3233.

12. Yamamoto, Y.; Takanashi, H.; Hettiarachchi, S.; Samarawickrama, S. Verification of the Destruction Mechanism of Structures in Sri Lanka and Thailand Due to the Indian Ocean Tsunami. *Coast. Eng. J.* **2006**, *48*, 117–145. [CrossRef]

13. Yamamoto, Y.; Charusrojthanadech, N.; Nariyoshi, K. Proposal of Rational evaluation method for damage to land structures due to tsunami. *J. JSCE B2 (Coast. Eng.)* **2011**, *67*, 72–91. (In Japanese)

14. Charusrojthanadech, N.; Yamamoto, Y.; Ca, V.T. Trial of Damage Evaluation by Indian Ocean Tsunami in Patong Beach of Thailand. *Coast. Eng. J.* **2011**, *53*, 285–317. [CrossRef]

15. Charusrojthanadech, N.; Yamamoto, Y.; Ca, V.T. Trial of an Evaluation Method of the Building Damage by Tsunami. *J. Earthq. Tsunami* **2012**, *6*, 31. [CrossRef]

16. Ribberink, J.S. Bed-load Transport for Steady Flows and Unsteady Oscillatory Flows. *Coast. Eng.* **1998**, *34*, 59–82. [CrossRef]

17. Yokoyama, Y.; Tanabe, H.; Torii, K.; Kato, F.; Yamamoto, Y.; Arimura, J.; Ca, V.T. A quantitative valuation of scour depth near coastal structures. In Proceedings of the International Conference on Coastal Engineering, Cardiff Wales, 7–12 July 2002; pp. 1830–1841.

18. Soulsby, R. Dynamics of Marine Sands. In *A Manual for Practical Application*; Thomas Telford: London, UK, 1997; 249p.

19. Zyserman, J.A.; Fredsoe, J. Data Analysis of Bed Concentration of Suspended Sediment. *J. Hydraul. Eng. ASCE* **1994**, *120*, 1021–1042. [CrossRef]

20. Iwanuma City. The Multiplex Protection Plan of Iwanuma City. 2015. Available online: https://www.city. iwanuma.miyagi.jp/bosai/bosai-bohan/bosai/documents/tunamihinannkeikaku.pdf and https://www.city. iwanuma.miyagi.jp/bosai/fukko/seibi/documents/grand.pdf (accessed on 10 May 2020). (In Japanese).

21. Port and Airport Research Institute. *Urgent Survey for 2011 Great East Japan Earthquake and Tsunami Disaster in Ports and Coasts*; Technical Note of PARI; Japan Research and Development Institute: Yokosuka, Kanagawa, Japan, 2011; 200p. (In Japanese)

22. Udo, K.; Sugawara, D.; Tanaka, H.; Imai, K.; Mano, A. Impact of the 2011 Tohoku Earthquake and Tsunami on Beach Morphology along the Northern Sendai Coast. *Coast. Eng. J.* **2012**, *54*, 1250009-1–1250009-15. [CrossRef]

23. Nanayakkara, K.I.U.; Dias, W.P.S. Fragility Curves for Structures under Tsunami Loading. *J. Nat. Hazards* **2016**, *80*, 471–486. [CrossRef]

24. Suppasri AMas, E.; Koshimura, S.; Imai, K.; Harada, K.; Imamura, F. Developing Tsunami Fragility Curves from the Surveyed Data of the 2011 Great East Japan Tsunami in Sendai and Ishinomaki Plains. *Coast. Eng. J.* **2012**, *54*, 1250008-1–1250008-16. [CrossRef]

25. Suppasri, A.; Mas, E.; Charvet, I.; Gunasekera, R.; Imai, K.; Yo, F.; Abe, Y.; Imamura, F. Building damage characteristics based on surveyed data and fragility curves of the 2011 Great East Japan tsunami. *J. Nat. Hazards* **2013**, *66*, 319–341. [CrossRef]
26. Suppasri, A.; Ingrid, C.; Kentaro, I.; Fumihiko, I. Fragility Curves Based on Data from the 2011 Tohoku-Oki Tsunami in Ishinomaki City, with Discussion of Parameters Influencing Building Damage. *J. Earthq. Spectra* **2015**, *31*, 841–868. [CrossRef]
27. Asakura, R.; Iwase, K.; Ikeya, T.; Takao, M.; Kaneto, T.; Fujii, N.; Omori, M. An experimental study on wave force acting on on-shore structures due to overflowing tsunamis. In Proceedings of the 27th Conference on Coastal Engineering, JSCE, Sydney, Australia, 16–21 July 2000; Volume 47, pp. 911–915. (In Japanese).
28. Iizuka, H.; Matsutomi, H. Damage due to the flooding flow of tsunami. In Proceedings of the 27th Conference on Coastal Engineering, JSCE, Sydney, Australia, 16–21 July 2000; Volume 47, pp. 381–385. (In Japanese).
29. Notification 1452 of the Ministry of Construction, Japan, Chiyoda-ku, Tokyo. 2000. Available online: www.house-support.net/tisiki/muku.htm (accessed on 1 September 2020).

Journal of
Marine Science and Engineering

Article

Applicability of Calculation Formulae of Impact Force by Tsunami Driftage

Yoshimichi Yamamoto [1,*], Yuji Kozono [2], Erick Mas [3], Fumiya Murase [4], Yoichi Nishioka [5], Takako Okinaga [6] and Masahide Takeda [7]

1. Department of Civil Engineering, Tokai University, Hiratsuka 259-1292, Kanagawa, Japan
2. Nippon Koei Co., Ltd., Tsukuba 300-1259, Ibaraki, Japan; kozono-yj@n-koei.jp
3. International Research Institute of Disaster Science, Tohoku University, Sendai 980-8572, Miyagi, Japan; mas@irides.tohoku.ac.jp
4. Shimizu Corporation, Sendai 980-0801, Miyagi, Japan; f.murase@shimz.co.jp
5. Kokusai Kogyo Co., Ltd., Amagasaki 660-0805, Hyogo, Japan; yoichi_nishioka@kk-grp.jp
6. Kajima Corporation, Chofu 182-0036, Tokyo, Japan; fukuytak@kajima.com
7. Research and Development Center, Toa Corporation, Yokohama 230-0035, Kanagawa, Japan; m_takeda@toa-const.co.jp
* Correspondence: yo-yamamo@tokai.ac.jp or yama1231@krd.biglobe.ne.jp

Citation: Yamamoto, Y.; Kozono, Y.; Mas, E.; Murase, F.; Nishioka, Y.; Okinaga, T.; Takeda, M. Applicability of Calculation Formulae of Impact Force by Tsunami Driftage. *J. Mar. Sci. Eng.* **2021**, *9*, 493. https://doi.org/10.3390/jmse9050493

Academic Editor: Eugen Rusu

Received: 31 December 2020
Accepted: 28 April 2021
Published: 1 May 2021

Publisher's Note: MDPI stays neutral with regard to jurisdictional claims in published maps and institutional affiliations.

Abstract: The aftermath of the Indian Ocean tsunami on 26 December 2004 triggered by the off Sumatra earthquake (magnitude "M" = 9.1), and the Great East Japan earthquake of 11 March 2011 off the Pacific coast of Tohoku (M = 9.0), evidence the secondary damage from driftage collision due to large tsunami waves. To prevent this type of damage, the establishment of methods for predicting driftage movement and calculating the impact force by driftage is necessary. Several numerical models have been developed to predict the driftage movement of objects. Every year, these improve in accuracy and usability. In contrast, there are many calculation formulae for calculating the impact force. However, since there are considerable differences between values calculated using these formulae, the reliability of each formula is unknown. Therefore, in this research, one team of the committee on tsunami research of the Japan Society of Civil Engineers summarizes the main calculation formulae of impact forces that have been proposed until 2019. In addition, for each type of driftage (driftwood, containers, cars, ships), we compare calculation values of these formulae with measured data of large-scale experiments. Finally, we check the range of calculation values for each formula up to 15 m/s in collision velocity and clarify then the following facts: (1) In the case of driftwood, the formulae of Matsutomi, Federal Emergency Management Agency (FEMA) and National Oceanic and Atmospheric Administration (NOAA), and American Society of Civil Engineers (ASCE) are most reliable; (2) In the case of containers, the formulae of Matsutomi, Arikawa et al., FEMA and NOAA, Ikeno et al., and ASCE are most reliable; (3) In the case of cars, the formulae of FEMA and NOAA, and ASCE are most reliable; (4) In the case of ships, the formulae of Mizutani, FEMA and NOAA, and ASCE are most reliable.

Keywords: car and ship; container; driftwood; impact force; large tsunami; tsunami driftage

1. Introduction

When a large tsunami caused by an earthquake reaches coastal communities, secondary damage by driftage and collision of objects is observed. For instance, when the damage surveys in Thailand (January 2005) and in Sri Lanka (May 2005) took place, following the off Sumatra earthquake (magnitude "M" = 9.1) and the Indian Ocean tsunami, this type of damage was confirmed. Later, damage surveys conducted in Japan (March and April 2011) in the aftermath of the Great East Japan earthquake (M = 9.0) evidenced, once again, the significant damage caused by driftage and collision of objects. Some examples are shown in the following.

(1) Driftwood

Figure 1a shows fallen trees that collided against a hotel at Khao Lak in Thailand, while Figure 1b portraits some logs that pierced the windows of a building at Ofunato in Japan.

Figure 1. Damage from driftwood: (**a**) fallen trees collided against a hotel at Khao Lak in Thailand, and (**b**) some logs pierced the windows of a building at Ofunato in Japan.

(2) Containers

In Figure 2a, a container collided against a house near the coast at Hitachi in Japan. Similarly, in Figure 2b, a container broke the wall of a large warehouse at the Soma port in Japan.

Figure 2. Damage from shipping containers: (**a**) a container collided against a house near the coast at Hitachi in Japan, and (**b**) a container broke the wall of a large warehouse at the Soma port in Japan.

(3) Cars

Figure 3a shows a dump truck that collided against a warehouse at the Kirinda fishing port in Sri Lanka. In Figure 3b, a passenger car collided against a shop at Miyako in Japan. Figure 3c shows how cars were pushed against the corner of a school building at Sendai in Japan.

Figure 3. *Cont.*

Figure 3. Damage from cars: (**a**) a dump truck collided against a warehouse at the Kirinda fishing port in Sri Lanka, (**b**) a passenger car collided against a shop at Miyako in Japan, and (**c**) many cars were pushed against the corner of a school building at Sendai in Japan.

(4) Ships

In Figure 4a, a patrol boat stopped at the foot of a mountain 1 km inland of Khao Lak in Thailand. In addition, Figure 4b shows a building with a big hole made by the patrol boat in Khao Lak. On the other hand, Figure 4c shows a large fishing boat left behind in the ruins near a fishing port at Kesennuma in Japan. Figure 4d shows a broken steel frame building on the shortest line between the large fishing boat and the fishing port.

Figure 4. Damage due to ships: (**a**) a patrol boat stopped at the foot of a mountain located 1 km inland of Khao Lak in Thailand, (**b**) a building showing a big hole made by the patrol boat in Khao Lak in Thailand, (**c**) a large fishing boat left behind in the ruins near a fishing port at Kesennuma in Japan, and (**d**) a broken steel frame building that was located on the shortest line between the large fishing boat and the fishing port at Kesennuma in Japan.

To prevent damage by driftage, it is important to establish a method for predicting driftage movement and calculating the impact force by driftage. Nistor et al. [1] reviewed existing papers published until 2016 about the transportation and the impact force of debris caused by a tsunami.

Experiential methods to predict driftage movement can be easily handled as the range at which driftage can reach is estimated based on simple assumptions, as demonstrated by Naito et al. [2]. In contrast, driftage movement can be evaluated through numerical simulations (Nistor et al. [1]), though the further improvement of the simulation accuracy and the convenience of the handling for engineers is desired. Regarding evaluation methods by numerical models, Yoneyama et al. [3] conducted a study by which the 3-dimensional VOF method was used for fluid motion calculation, and Murase et al. [4] used the storm surge and tsunami simulator in oceans and coastal areas (STOC) model.

For methods to calculate the impact force, although there are numerical models, such as the elastoplastic finite deformation analysis of Magoshi et al. [5], they have not reached the level at which they can be used simply and reliable for engineers. Thus, traditional methods generally used are semi-theoretical formulae and empirical formulae, as shown in Tables 1–3. However, since there are considerable differences between values calculated by these formulae, the reliability of each formula is unknown. In this regard, Kaida and Kihara [6] investigated calculation formulae of impact forces proposed until 2012, and they checked the performance range of calculated values by each formula up to a maximum value of collision velocity of 10 m/s. Finally, they summarized the calculation formulae that are effective for each type of driftage. Moreover, they compiled the appropriate values of shaft rigidity, which is an important parameter in some formulas presented in Tables 1–3. Furthermore, Stolle et al. [7,8] indicated that the calculation accuracy of the impact force of low rigid debris (like driftwood, containers) could be improved by estimating the rigidity of the debris exactly.

Table 1. Impact force formula (no. 1).

Driftage	Formula	Reference
Driftwood	$$F_i = u_{max} \sqrt{k(m_d + c_M m_f)} \times \beta e$$ $$\beta = \sin \varphi, e = \cfrac{1}{\sqrt{1 + \left[\left(\frac{\varepsilon_o}{r_i} \right) \left(1 + \mu \frac{r_o}{\varepsilon_o} \right) \right]}}$$ F_i is the impact force (N); u_{max} is the maximum velocity of driftwood; k is the effective contact stiffness; m_d is the mass of driftwood; c_M is the added mass coefficient; m_f is the mass of displaced fluid; β is the decreasing coefficient by the orientation angle φ; e is the decreasing coefficient by eccentricity; ε_o is the length from the barycenter to the collided point; r_i is the turning radius; μ is the friction coefficient between the driftwood and the collided body; r_o is the radius of the sectional area of the driftwood.	Haehnel and Daly (2004) [9]
Driftwood and Containers	$$F_i = \frac{\pi m u c_I c_o c_D c_B R_{max}}{2 \Delta t}, \Delta t = \frac{\pi}{2} \sqrt{\frac{m}{k}}$$ F_i is the impact force (N); m is the mass of the driftage; u is the collision velocity; c_I is the important factor; c_o is the orientation coefficient; c_D is the coefficient of water depth; c_B is the barrier coefficient; R_{max} is the maximum response ratio to the collision load; Δt is the continuing time of the collision; k is the effective stiffness.	ASCE (2010) [10]
Driftwood (the Mass is about 450 kg) and Containers (the Empty Mass is 2200–3800 kg)	$$F_i = 1.3 u_{max} \sqrt{m_d k (1 + c)}, u_{max} = \sqrt{2gR\left(1 - \frac{z}{R}\right)}$$ F_i is the impact force (N); 1.3 is the importance coefficient of risk category IV; u_{max} is the maximum fluid velocity; m_d is the mass of driftage; k is the effective stiffness (2.4×10^6 N/m to driftwood; 6.5×10^8 N/m to 12.2 m empty containers; 1.5×10^9 N/m to 6.1 m empty containers); c is the hydrodynamic mass coefficient (0.0 to the long axis direction for the collision of driftwood; 0.2 to the long axis direction for the collision of 12.2 m containers; 1.0 to the transverse direction for the collision of 12.2 m containers; 0.3 to the long axis direction for the collision of 6.1 m containers; 1.0 to the transverse direction of 6.1 m containers); g is the gravitational acceleration; R is the maximum run-up height $\times$ 1.3; z is the ground height on the water surface.	FEMA & NOAA (2012) [11]

Table 2. Impact force formula (no. 2).

Driftage	Formula	Reference
Driftwood, Containers, etc.	$F_i = u\sqrt{k_d m_d}, k_d = \frac{EA_d}{L_d}$ F_i is the impact force (N), u is the collision velocity; k_d and m_d are the stiffness and the mass of driftage; E is the elastic modulus of driftage; A_d and L_d are the sectional area and the length of driftage. The impulse duration of the steel tube can be obtained using $t_d = 2\sqrt{\frac{m_d}{k_d}}$	Aghl, Naito, and Riggs (2014) [12]
Driftwood, Containers (6.1–12.2 m), Cars, Ships, etc.	$F_i = R_{max} I_{TSU} C_o u_{max} \sqrt{km_d}$ F_i is the impact force (N); R_{max} is the dynamic response ratio to the collision load (it is in the range of 0.0–1.8 according to the collision action time); I_{TSU} is the importance factor (1.0–1.25); C_o is the orientation coefficient (=0.65); u_{max} is the maximum fluid velocity; k is the smaller one of (i) the stiffness of the driftage; (ii) the transverse direction stiffness of the collided body; m_d is the mass of the driftage. The stiffness of the driftwood can be calculated by using $k = EA/L$. E is the elasticity modulus of the long axis direction of driftage; A and L are the sectional area and the length. The impulse duration for empty containers t_d can be calculated by using $t_d = 2\sqrt{\frac{m_d}{k}}$ The impulse duration for loaded containers t_d can be calculated by using $t_d = \frac{m_d + m_{contents}}{\sqrt{km_d}}$ Here, $m_{contents}$ is 50% of the maximum capacity of the container.	ASCE (2015) [13]
Ships	$F = \frac{WV^2}{4gl_d}$ F is the impact force (kN); W is the weight of the ship (kN); V is the collision velocity; l_d is the stopping distance. Table of stopping distance of the ship 	Ship line	Weight of the ship (ton-force)	Stopping distance (m)
Main line	200,000	8
Side line	5000	1		Bridge and Engineering Association (1978) [14]
Driftwood (Long Axis Direction Collision)	$\frac{F}{\gamma D^2 L} = 1.6\, C_{MA}\left\{\frac{V}{(gD)^{0.5}}\right\}^{1.2}\left(\frac{\sigma_f}{\gamma L}\right)^{0.4}$ F is the impact force (kN); C_{MA} is the apparent mass coefficient (1.7 for a bore or a surge, 1.9 for a steady flow); V is the collision velocity (m/s); D and L are the diameter and the length of driftwood (m); σ_f is the yield stress of driftwood (kN/m^2); γ is the unit weight of driftwood (kN/m^3); g is the gravitational acceleration.	Matsutomi (1999) [15]

Table 3. Impact force formula (no. 3).

Driftage	Formula	Reference
Driftwood, Cars, Ships, etc.	$F = 0.1 \cdot W \cdot v$ F is the impact force (kN): W is the weight of driftage (kN); v is the surface fluid velocity (m/s).	Japan Road Association (2002) [16]
Driftwood	$\frac{F}{gM} = S \cdot C_{MA}\left\{\frac{V}{g^{0.5} D^{0.25} L^{0.25}}\right\}^{2.5}$ F is the impact force (kN); S is a coefficient (5.0); C_{MA} is the added mass coefficient [pillar oriented transversely: 2.0 (two dimension), 1.5 (three dimension); square pillar oriented transversely: 2.0–4.0 (two dimension); 1.5 (three dimension); pillar oriented longitudinally: about 2.0, sphere: about 0.8]; V is the bore fluid velocity (m/s); D is the collision section size (m); L is the length of driftwood (m); M is the mass of driftwood (kg); g is the gravitational acceleration (m/s^2).	Ikeno and Tanaka (2003) [17]

Table 3. *Cont.*

Driftage	Formula	Reference
Containers	$$F = 2\rho_w \eta_m B_c V_x^2 + \frac{WV_x}{gdt}$$ F is the impact force (kN); ρ_w is the water density (t/m^3); η_m is the maximum run-up height; B_c is the width of the container; V_X is the drift velocity (m/s); W is the container weight; g is the gravitational acceleration; dt is the collision time.	Mizutani et al. (2005) [18]
Small Ships	$$F = 2\frac{WV_x}{gdt}$$ F is the impact force (kN); W is the weight of the small ship (kN); V_X is the drifting velocity of the small ship (m/s); g is the gravitational acceleration; dt is the collision time.	Mizutani et al. (2007) [19]
Driftwood, Containers, etc.	$$F = \gamma_p \chi^{2/5}\left(\tfrac{5}{4}\tilde{m}\right)^{3/5} V^{6/5}$$ $$\chi = \frac{4\sqrt{a}}{3\pi}\frac{1}{k_1+k_2}, \; k_i = \frac{1-v_i^2}{\pi E_i},$$ for the driftwood: $\tilde{m} = C_{MA}m_1$ for the containers: $\tilde{m} = \frac{m_1 m_2}{m_1+m_2}$ F is the impact force (kN); a is 1/2 of the radius of the collision sectional area (m); E is the elastic modulus (kN/m^2]; v_i is Poisson's ratio; m_i is the mass (kg); V is the collision velocity (m/s); γ_p is the energy attenuation effect by plasticity (0.25); the suffix i of m and k indicates the driftage and the collided body.	Arikawa et al. (2007) [20], (2010) [21]
Driftwood and Containers	$$F = k_c (C_{MA}M)^{0.6} V^{1.2} D^{0.2} E^{0.4}$$ F is the impact force at the time of the elastic collision (kN); k_c is a constant 0.243; C_{MA} is the apparent mass coefficient (2.0 for wooden pillar oriented longitudinally, 1.0 for containers); M is the mass of driftage (kg); V is the collision velocity (m/s); D is the collision sectional area size (m); E is the representative stiffness $= \dfrac{E_1 \times E_2}{E_1\left(1-v_2^2\right)+E_2\left(1-v_1^2\right)}$	Ikeno et al. (2013) [22]

In this research, we summarize the main calculation formulae of impact forces proposed by 2019, and the calculation values by these formulae are compared with measured data of large-scale experiments for each type of driftage (driftwood, shipping containers, cars, ships). Moreover, the range of calculated values by each formula up to a collision velocity of 15 m/s is investigated. Finally, the reliability of each formula is discussed and clarified.

2. Accuracy of Formulae for Calculating the Impact Force

2.1. List of Main Calculation Formulae

The main calculation formulae of impact forces by tsunami driftage proposed by 2019 are summarized in Tables 1–3. Although these calculation formulae were designed to reflect the proposer's experimental data, there are considerable differences among the proposed formulae. A possible reason for these differences might be that several formulae cannot make the difference in rigidity between the driftage and the collided body correctly into consideration. When the driftage collides with the collided body, the impact force is reduced by carrying out an elastic modification. Therefore, the impact force is different if the rigidity of the driftage or the collided body is not identical, even when the velocity at the time of collision is the same. In other words, if the differences of bodies with different rigidity from the experiment used to verify the proposed formula are not considered, the accuracy of the results is reduced.

In the case of the tsunami force in which seawater collides against a rigid body made of concrete or steel, the seawater deforms easily compared with the collided body. Moreover, the inverse of whole rigidity is proportional to the inverse of each rigidity of the collision body and the collided body. Therefore, we can say that the rigidity of the seawater is very low, and the decrease of the tsunami force is governed not by the collided body but by seawater. Therefore, we do not need to make the difference in rigidity of the collided body into consideration.

2.2. Correlation Examination of the Main Calculation Formulae

We searched for large-scale experimental data with the size, weight, rigidity of the driftage and the collided body, the collision velocity, and the impact force for each driftage type (driftage, containers, cars, ships) in the published literature. We compare the calculated values of the impact force and measurements.

Here, as the improved formula of Haehnel and Daly [9] is the formula of FEMA and NOAA [11] and the improved formula of ASCE [10] and Aghl, Naito and Riggs [12] is the formula of ASCE [13], the formulae of Haehnel and Daly [9], ASCE [10], and Aghl, Naito and Riggs [12] are not examined.

Because we could not find materials, which can determine the stop distance of driftage except for ships, the formula of the Bridge and Engineering Association [14] was examined only for the case of ships. Moreover, since we could not find materials, which can determine the yield stress of ships, the formula of Matsutomi [15] was not examined for the case of ships. The formula of Mizutani et al. [18] was examined for containers and cars, which can estimate a maximum run-up height easily because they do not float and move immediately. In addition, the formula of [19] was examined in the case of ships. Moreover, for the formula of ASCE [13], the values for R_{max}, I_{TSU}, C_o were set as 1.8, 1.25, and 0.65, respectively. Although ASCE [13] recommends some values for the effective stiffness k, when this was not available, the recommended values by FEMA and NOAA [11] were used.

2.2.1. Driftwood

Although there are many small-scale experiments, we selected the large-scale experiments of Matsutomi [15] and Ikeno et al. (Central Research Institute of Electric Power Industry) [23] because they specify necessary information, and we actually used the data for collision velocities larger than 2.0 m/s in their experiments, which were measured in unsteady flow using a water flume.

The calculation conditions are as follows:

(1)   Experiment 1 of Matsutomi [15]

   (a)   Parameters of the driftwood (larch)

   0.3 m in diameter, 2.1 m long, 78.67 kg, 25 MN/m^2 yield stress, 0.4 Poisson's ratio, 8000 MN/m^2 elasticity modulus.

   (b)   Parameters of the collided body (steel)

   0.3 Poisson's ratio, 200,000 MN/m^2 elasticity modulus.

   (c)   Coefficients in the formula

   Each apparent mass coefficient C_{MA} was set up according to each proposer's recommended value (1.7 for Matsutomi [15], 2.0 for Ikeno and Tanaka [17], 1.7 for Arikawa et al. [20], 2.0 for Ikeno et al. [22]). The constant 0.0 was set up to the coefficient C of FEMA and NOAA [11]. The axial stiffness k_1 of the driftwood was 269 MN/m (since the axial stiffness changes according to the dimensions of the object, the recommended value of FEMA and NOAA [11] is usable only to specified dimensions. Therefore, this value was calculated using the elasticity modulus of driftwood), and the bending stiffness k_2 of the steel was 5 MN/m.

   (d)   Measured impact force and the collision velocity

   The measured impact force and the collision velocity are 0.022 MN and 2.0 m/s, respectively.

(2)   Experiment 2 of Matsutomi [15]

   (a)   Parameters of the driftwood (larch)

   0.3 m in diameter, 2.7 m long, 101.15 kg, 25 MN/m^2 yield stress, 0.4 Poisson's ratio, 8000 MN/m^2 elasticity modulus.

   (b)   Parameters of the collided body (steel) and the apparent mass coefficients used the same values as (1) Matsutomi's experiment 1 [15].

   (c)   Coefficients in the formula;

The coefficient C of FEMA and NOAA [11] was chosen as 0.0. The axial stiffness k_1 of the driftwood was 209 MN/m, and the bending stiffness k_2 of the steel was 5 MN/m.

(d)   Measured impact force and the collision velocity

The measured impact force and the collision velocity are 0.026 MN and 2.0 m/s, respectively.

(3)   Matsutomi's experiment 3 [15]

(a)   Parameters of the driftwood (larch)

0.3 m in diameter, 4.0 m long, 149.85 kg, 25 MN/m^2 yield stress, 0.4 Poisson's ratio, 8000 MN/m^2 elasticity modulus

(b)   Parameters of the collided body (steel) and the apparent mass coefficients used the same values as (1) Matsutomi's experiment 1 [15].

(c)   Coefficients in the formula;

The coefficient C of FEMA and NOAA [11] was 0.0. The axial stiffness k_1 of the driftwood was 141 MN/m, and the bending stiffness k_2 of the steel was 5 MN/m.

(d)   Measured impact force and the collision velocity

The measured impact force and the collision velocity are 0.036 MN and 2.0 m/s, respectively.

(4)   Experiment 1 of Ikeno et al. [23];

(a)   Parameters of the driftwood (pine);

0.42 m in diameter, 2.0 m long, 177.0 kg, 25 MN/m^2 yield stress, 0.4 Poisson's ratio, 11,200 MN/m^2 elasticity modulus.

(b)   Parameters of the collided body (steel) and the apparent mass coefficients used the same values as (1) Matsutomi's experiment 1 [15].

(c)   Coefficients in the formula;

The coefficient C of FEMA and NOAA [11] was 0.0. The axial stiffness k_1 of the driftwood was 776 MN/m, and the bending stiffness k_2 of the steel was 12.8 MN/m.

(d)   Measured impact force and the collision velocity;

The measured impact force and the collision velocities are 0.080 MN and 2.0 m/s, 0.082 MN and 4.0 m/s, respectively.

(5)   Experiment 2 of Ikeno et al. [23]

(a)   Parameters of the driftwood (cedar)

0.382 m in diameter, 1.0 m long, 73.0 kg, 23 MN/m^2 yield stress, 0.4 Poisson's ratio, 9100 MN/m^2 elasticity modulus.

(b)   Parameters of the collided body (steel) and the apparent mass coefficients used the same values as (1) Matsutomi's experiment 1 [15].

(c)   Coefficients in the formula;

The coefficient C of FEMA and NOAA [11] was 0.0. The axial stiffness k_1 of the driftwood was 1040 MN/m, and the bending stiffness k_2 of the steel was 12.8 MN/m.

(d)   Measured impact force and the collision velocity

The measured impact force and the collision velocity are 0.054 MN and 4.1 m/s, respectively.

(6)   Experiment 3 of Ikeno et al. [23];

(a)   Parameters of the driftwood (cedar)

0.382 m in diameter, 2.0 m long, 156.0 kg, 23 MN/m^2 yield stress, 0.4 Poisson's ratio, 9100 MN/m^2 elasticity modulus.

(b)   Parameters of the collided body (steel) and the apparent mass coefficients used the same values as (1) Matsutomi's experiment 1 [15].

(c)   Coefficients in the formula

The coefficient C of FEMA and NOAA [11] was 0.0. The axial stiffness k_1 of the driftwood was 521 MN/m, and the bending stiffness k_2 of the steel was 12.8 MN/m.

(d)   Measured impact force and the collision velocity

The measured impact force and the collision velocity are 0.077 MN and 4.1 m/s, respectively. The verification results are shown in Figure 5.

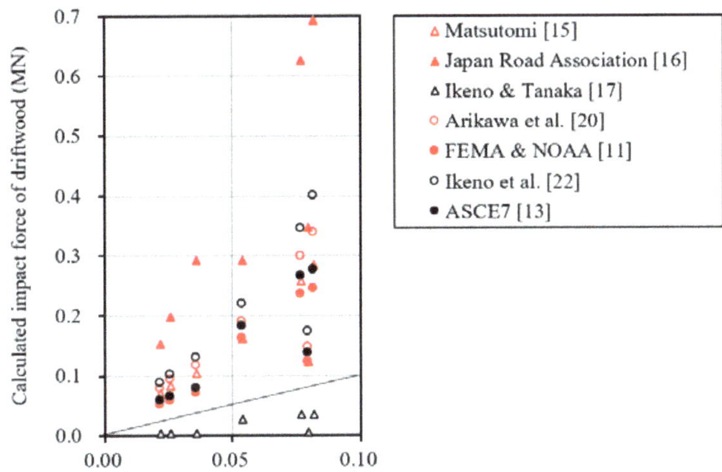

Figure 5. Comparison of impact forces for driftwood.

The Japan Road Association's formula [16] overestimates, and Ikeno and Tanaka's formula [17] underestimates the measured data. The formulae of FEMA and NOAA [11], ASCE [13], and Matsutomi [15] result in values reasonably close to the measured values. Since the formulae of Arikawa et al. [20] and Ikeno et al. [22] correlate well with measured values, provided the formulae are improved so that parameters concerning the rigidity of the driftage and the collided body can be decided adequately, the calculation accuracy of the improved formulae will improve.

2.2.2. Containers

Large-scale experiments with the necessary information were conducted by Aghl et al. [12] and Arikawa et al. [20] (Port and Airport Research Institute). Only data with collision velocities larger than 1.4 m/s were used. The experiment of Aghl et al. [12] was implemented in the air by using a full-scale container, and the experiment of Arikawa et al. [20] was conducted in unsteady flow using a water flume and a container model of 1:5 in scale. Moreover, when using the formula of Mizutani et al. [18], 1000 kg/m^3 was used for ρ_w since freshwater was used in their experiment, and half of the height of their containers was used for η_m so that $2 \times \eta_m \times B_c$ becomes a collision sectional area. Furthermore, the impulse duration of ASCE [13] was used for dt.

The calculation conditions are as follows:

(1)   Experiment of Arikawa et al. [20]

(a)   Parameters of the small container (steel)

0.5 m high, 1.21 m long, 62.00 kg, 250 MN/m^2 yield stress, 0.3 Poisson's ratio, 200,000 MN/m^2 elasticity modulus.

(b)   Parameters of the collided body (concrete)

0.2 Poisson's ratio, 25,000 MN/m^2 elasticity modulus.

(c)   Coefficients in the formula

Each apparent mass coefficient C_{MA} was set up according to each proposer's recommendation value (1.7 for Matsutomi [15], 4.0 for Ikeno and Tanaka [17], 1.0 for Ikeno et al. [22]). The coefficient C of FEMA and NOAA [11] was 0.2. The axial stiffness k_1 of the small container was 1080 MN/m (since the axial stiffness changes according to the dimensions of the object, the recommended value of FEMA and NOAA [11] cannot be used. Therefore, this value was calculated using the elasticity modulus of the small container), and the bending stiffness k_2 of the concrete was 2220 MN/m.

(d)   Measured impact force and the collision velocity

The measured impact force and the collision velocities are 0.033 MN and 1.70 m/s, 0.049 MN and 2.15 m/s, 0.057 MN and 2.40 m/s, respectively.

(2)   Experiment of Aghl et al. [12]

(a)   Parameters of the 6.1 m container (steel)

2.5 m high, 6.1 m long, 2300 kg, 250 MN/m^2 yield stress, 0.3 Poisson's ratio, 200,000 MN/m^2 elasticity modulus.

(b)   Parameters of the collided body (steel)

0.3 Poisson's ratio, 200,000 MN/m^2 elasticity modulus.

(c)   Coefficients in the formula

Each apparent mass coefficient C_{MA} was set up as the ones in the air (1.0 for Matsutomi [15], 2.0 for Ikeno and Tanaka [17], 1.0 for Ikeno et al. [22]). The constant 0.0 and 85 MN/m, which FEMA and NOAA [11] recommends, were set up for the coefficient C and the effective stiffness k of FEMA and NOAA [11], respectively. Moreover, 42.9 MN/m recommended by ASCE [13] was set up for the effective stiffness k of ASCE [13].

(d)   Measured impact force and the collision velocity

The measured impact force and the collision velocities are 0.795 MN and 1.43 m/s, 0.995 MN and 1.80 m/s, 1.195 MN and 2.13 m/s, 1.315 MN and 2.42 m/s, respectively.

The verification results are shown in Figure 6.

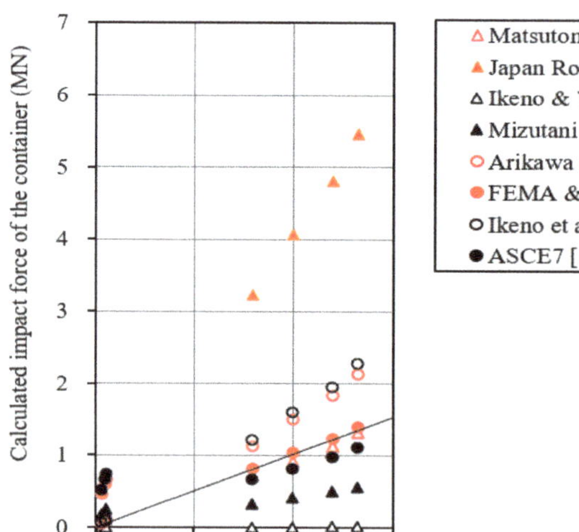

Figure 6. Comparison of impact forces for the containers.

The formula of the Japan Road Association [16] overestimates, and the formula of Ikeno and Tanaka [17] underestimates the measured data. The formulae of FEMA and

NOAA [11], ASCE [13], and Matsutomi [15] provide values near the measured values. Since the formulae of Mizutani et al. [18], Arikawa et al. [20], and Ikeno et al. [22] show good correlations with the measured values, provided the formulae are improved so that parameters concerning rigidity of the driftage and the collided body can be decided adequately, the calculation accuracy of the improved formulae will improve.

2.2.3. Car

Unique large-scale experiments with all necessary information were conducted by Takabatake et al. [24] (Central Research Institute of Electric Power Industry). On the other hand, when using the formula of Mizutani et al. [18], 1000 kg/m^3 was used for ρ_w, and half of the height of their car was used for η_m. Furthermore, the impulse duration of ASCE [13] was used for dt.

The calculation conditions are as follows:

(a)   Parameters of the actual car

1.48 m high, 1.40 m wide, 3.30 m long, 316.0 kg, 83,700 N/m^2 yield stress (from data when the buckling occurred in Takabatake et al. [24]), 0.3 Poisson's ratio. The elasticity modulus was calculated by using the empirical equation of the axial stiffness (kN/m) (=457.41 × collision velocity-170.79) obtained from measured data of Takabatake et al. [24].

(b)   Parameters of the collided body (steel)

0.3 Poisson's ratio, 200,000 MN/m^2 elasticity modulus.

(c)   Coefficients in the formula

Each apparent mass coefficient C_{MA} was set up according to each proposer's recommended value (1.7 for Matsutomi [15], 2.0 for Ikeno and Tanaka [17], 2.0 for Ikeno et al. [22]). The coefficient C of FEMA and NOAA [11] was 0.4. The effective stiffness k was set up using the empirical equation obtained from measured data of Takabatake et al. [24].

(d)   Measured impact force and the collision velocity

The measured impact force and the collision velocities are 0.003 MN and 0.39 m/s, 0.024 MN and 0.98 m/s, 0.040 MN and 1.78 m/s, 0.042 MN and 1.86 m/s, 0.053 MN and 2.07 m/s, 0.039 MN and 2.12 m/s, 0.087 MN and 3.90 m/s, respectively.

The verification results are shown in Figure 7.

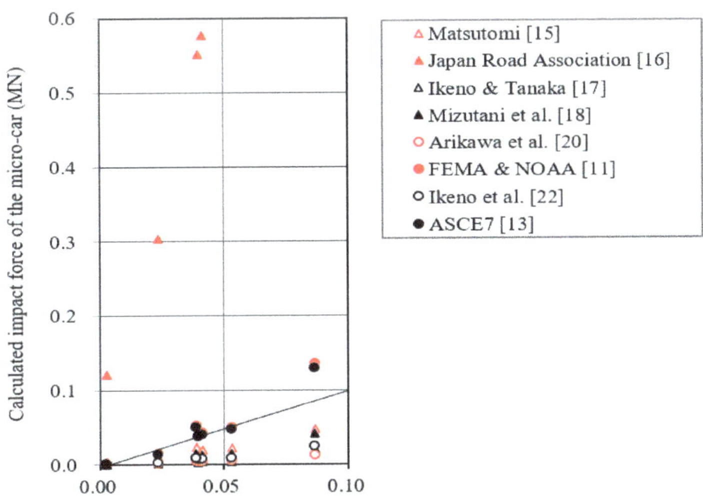

Figure 7. Comparison of impact forces for the microcar.

The formula of the Japan Road Association [16] overestimates the experimental values. The formulae of FEMA and NOAA [11] and ASCE [13] result in values near the measurements. Although the formulae of Matsutomi [15], Ikeno and Tanaka [17], Mizutani et al. [18], Arikawa et al. [20], and Ikeno et al. [22] underestimate the measurements, their predictions show a good correlation with measured values. Therefore, provided their formulae are improved so that parameters concerning the rigidity of the driftage and the collided body can be decided adequately, the calculation accuracy of the improved formulae will improve.

2.2.4. Ship

The large-scale experiments of Arita [25] (National Maritime Research Institute) were the only ones with all the necessary information. Even though in his experiment, a full-scale steel bow model collides against a steel pier model, it was also implemented in air, and the information, which we can use, is only a semi-theoretical formula based on data of this experiment. Moreover, since this formula consists only of the gross tonnage and the collision speed, this formula can only be applied when the conditions of the driftage and the collided body are in agreement with those of the experiment (other similar formulae cannot be used as general-purpose formulae owing to the same reason). When using the formula of Mizutani et al. [19], the impulse duration of ASCE [13] was used for dt.

The calculation conditions are as follows:

(a)   Parameters of the ship (500 GTs)

9.6 m wide, 50.0 m long, 1,500,000 kg, 0.51 m in stop distance (from the recommended values of the Bridge and Engineering Association [14]), 0.3 Poisson's ratio, 5,308,660 N/m in axial stiffness (from the data table for the axial stiffness of Takabatake et al. [24]), 5,760,000 N/m^2 elasticity modulus calculated from the axial stiffness by inverse analysis.

(b)   Parameters of the collided body (steel)

0.3 Poisson's ratio, 200,000 MN/m^2 elasticity modulus.

(c)   Coefficients in the formula

Each apparent mass coefficient C_{MA} was set up according to each proposer's recommended value (1.5 for Ikeno and Tanaka [17], 1.0 for Ikeno et al. [22]). The coefficient C of FEMA and NOAA [11] was 0.4. The effective stiffness k was set up using the data table for the axial stiffness of Takabatake et al. [24].

(d)   Measured impact force and the collision velocity

The measured impact force and the collision velocities are 1.96 MN and 0.8 m/s, 2.94 MN and 1.2 m/s, 3.92 MN and 1.6 m/s, 4.90 MN and 2.0 m/s, 5.88 MN and 2.4 m/s, respectively.

The formula of Ikeno and Tanaka [17] underestimates the experimental values.!The formula of Mizutani et al. [19] results in values very close to the experimental ones.

Although the formulae of FEMA and NOAA [11] and ASCE [13] overestimate the experimental values, their formulae have given values close to the experimental ones.

Although the formulae of the Bridge and Engineering Association [14], Japan Road Association [16], Arikawa et al. [20], and Ikeno et al. [22] underestimate the measured values, their formulae show good correlations with the experimental values. Therefore, provided their formulae are improved so that parameters concerning the rigidity of the driftage and the collided body can be decided adequately, the calculation accuracy of the improved formulae will improve.

The reason why the calculated values of each formula in Figure 5 are not aligned along the straight line as in Figures 6–8 is because the accuracies of the measured values are low since wood is a natural material and the collision angle of wood relative to the collided body is a sensitive parameter for the result. Moreover, the elasticity modulus and the stiffness of wood also vary significantly. In other cases, the materials of the driftage were fixed, and especially in the case of the full-scale air experiments, since collision angles were right-angled relatively to the collided body, there is no difference due to the collision angles in the measured impact forces.

The verification results are shown in Figure 8.

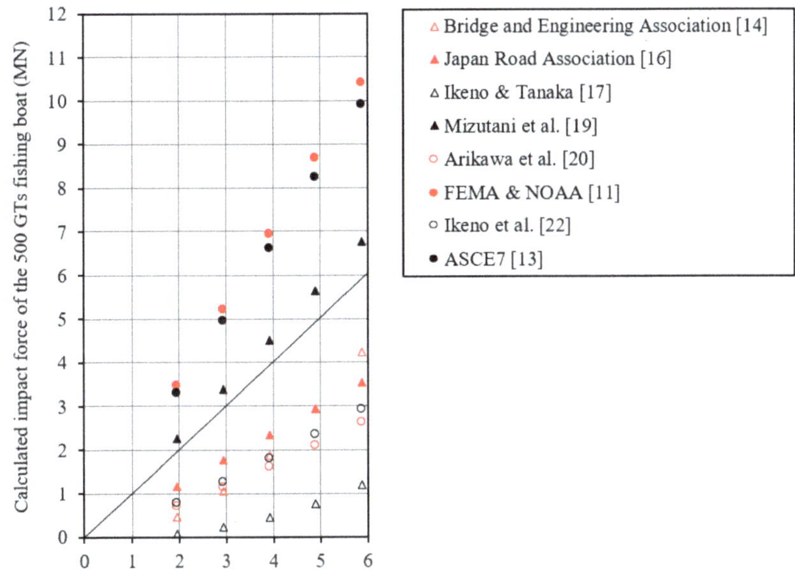

Figure 8. Comparison of impact forces for the ship.

2.3. Range of Impact Force of Each Formula

The collision velocity of the experimental data used for comparison with the previous section was not fast (0.39 m/s–4.1 m/s), and the range was small. Therefore, to check the validity of the calculation range of each formula, in the collision velocity range of 1 m/s–15 m/s, a general size was set up for each driftage type, and the impact force was calculated.

In addition, when the specific gravity of the driftage is larger than 1, the collision velocity must be lower than the tsunami velocity. Moreover, when a tsunami flow collides with a body with a large vertical plane, a backward flow occurs near the body. Therefore, as the driftage approaches the body, the collision velocity must become small by the backward flow. However, these influences are not considered.

2.3.1. Driftwood

(a)   Parameters of a slender wood (pine and cedar);

0.12 m in diameter, 1.6 m long, 11.56 kg, 34.3 MN/m^2 yield stress, 0.4 Poisson's ratio, 10,000 MN/m^2 elasticity modulus.

(b)   Parameters of a thick wood (pine and cedar);

0.51 m in diameter, 1.33 m long, 173.6 kg, 34.3 MN/m^2 yield stress, 0.4 Poisson's ratio, 10,000 MN/m^2 elasticity modulus.

(c)   Parameters of the collided body (concrete);

0.167 Poisson's ratio, 25,000 MN/m^2 elasticity modulus.

(d)   Coefficients in the formula;

Each apparent mass coefficient C_{MA} was set up according to each proposer's recommended value. (1.7 for Matsutomi [15], 2.0 for Ikeno and Tanaka [17], 1.7 for Arikawa et al. [20],

2.0 for Ikeno et al. [22]). The coefficient C of FEMA and NOAA [11] was 0.0. The effective stiffness k of the slender wood was 20.1 MN/m (this value was calculated using the elasticity modulus of the slender wood), and the effective stiffness k of the thick wood was 27.5 MN/m.

The results are shown in Figure 9. The formula of Ikeno and Tanaka [17] gives too small values compared with the mean values of all formulae.

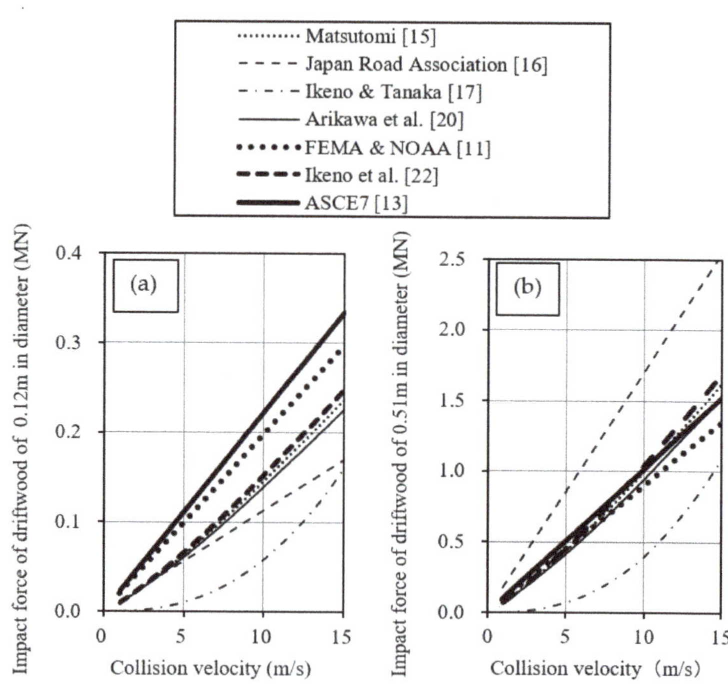

Figure 9. Impact force of driftwood: (**a**) the case of the slender wood, and (**b**) the case of the thick wood.

2.3.2. Containers

(a)   Parameters of an empty 12.2 m container;

2.45 m high, 12.1 m long, 3800 kg, 250 MN/m^2 yield stress, 0.3 Poisson's ratio, 200,000 MN/m^2 elasticity modulus.

(b)   Parameters of a fully-loaded 12.2 m container;

2.45 m high, 12.1 m long, 72,600 kg, 250 MN/m^2 yield stress, 0.3 Poisson's ratio, 200,000 MN/m^2 elasticity modulus.

(c)   Parameters of the collided body (concrete);

0.167 Poisson's ratio, 25,000 MN/m^2 elasticity modulus.

(d)   Coefficients in the formula;

Each apparent mass coefficient C_{MA} was set up according to each proposer's recommended value (1.7 for Matsutomi [15], 4.0 for Ikeno and Tanaka [17], 1.0 for Ikeno et al. [22]). The constant 0.2 and 60 MN/m recommended by FEMA and NOAA [11] were set up for the coefficient C and the effective stiffness k of FEMA and NOAA [11], respectively. Moreover, 29.8 MN/m recommended by ASCE [13] was set up for the effective stiffness k of ASCE [13].

When using the formula of Mizutani et al. [18], the density ρ_w was 1000 kg/m^3 and half of the height of their containers was used for η_m. Furthermore, the impulse duration of ASCE [13] was used for dt.

The results are shown in Figure 10. The formula of the Japan Road Association [16] gives too big values as compared with the mean values of all formulae.

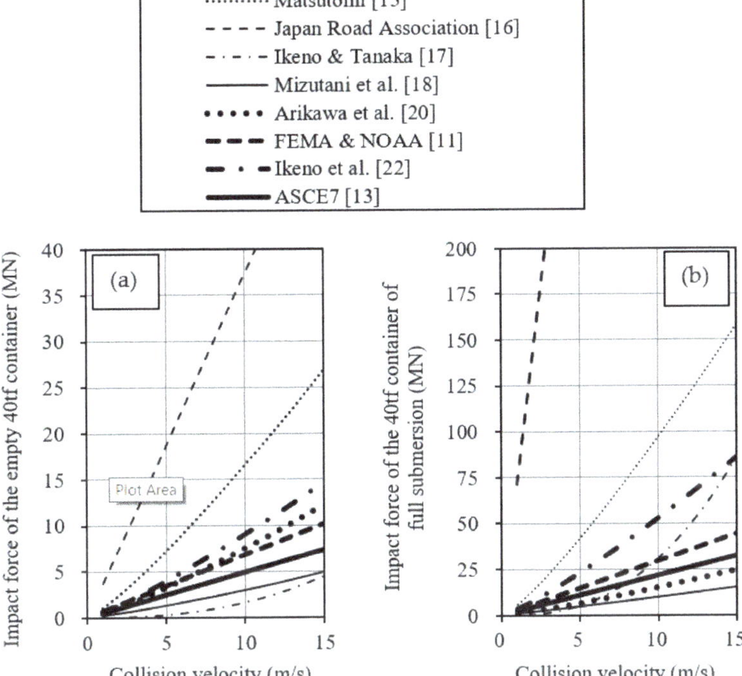

Figure 10. Impact force of the 12.2 m container: (**a**) the empty container, and (**b**) the fully-loaded container.

2.3.3. Car

(a)   Parameters of an actual car

1.80 m high, 2.00 m wide, 5.00 m long, 2500 kg, 47,800 N/m^2 yield stress (from data when the buckling occurred in Takabatake et al. [24]), 0.3 Poisson's ratio. The elastic modulus was calculated by using the empirical equation of the axial stiffness (kN/m) (=457.41 × collision velocity—170.79) obtained from measured data of Takabatake et al. [24].

(b)   Parameters of the collided body (concrete)

0.2 Poisson's ratio, 25,000 MN/m^2 elasticity modulus.

(c)   Coefficients of the formula

Each apparent mass coefficient C_{MA} was set up according to each proposer's recommendation value (1.7 for Matsutomi [15], 2.0 for Ikeno and Tanaka [17], 2.0 for Ikeno et al. [22]). The coefficient C of FEMA and NOAA [11] was 0.4. The effective stiffness k was set up using the empirical equation obtained from measured data of Takabatake et al. [24].

Moreover, when using the formula of Mizutani et al. [18], the density ρ_w was 1000 kg/m^3, and half of the height of the car was used for η_m. Furthermore, the impulse duration of ASCE [13] was used for dt.

The calculation results are shown in Figure 11. The formula of the Japan Road Association [16] gives too big values as compared with the mean values of all formulae.

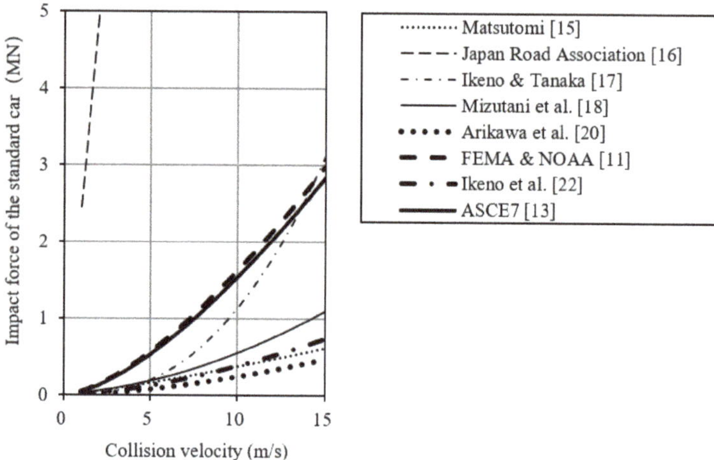

Figure 11. Impact force of the standard car.

2.3.4. Ships

(a)   Parameters of a 5 GT ship

2.8 m wide, 11.0 m long, 15,000 kg, 0.04 m in stop distance (from recommended values of the Bridge and Engineering Association [14]), 0.3 Poisson's ratio, 1144,950 N/m in axial stiffness (from the data table for the axial stiffness of Takabatake et al. [24]), 3,210,000 N/m^2 elasticity modulus calculated from the axial stiffness by inverse analysis.

(b)   Parameters of a 500 GT ship

9.4 m wide, 55.0 m long, 1,500,000 kg, 0.51 m in stop distance (from recommended values of the Bridge and Engineering Association [14]), 0.3 Poisson's ratio, 5,308,660 N/m in axial stiffness (from the data table for the axial stiffness of Takabatake et al. [24]), 6,610,000 N/m^2 elasticity modulus calculated from the axial stiffness by inverse analysis.

(c)   Parameters of the collided body (concrete)

0.2 Poisson's ratio, 25,000 MN/m^2 elasticity modulus.

(d)   Coefficients in the formula

Each apparent mass coefficient C_{MA} was set up according to each proposer's recommended value (1.5 for Ikeno and Tanaka [17], 1.0 for Ikeno et al. [22]). The coefficient C of FEMA and NOAA [11] was 0.4. The effective stiffness k was set up using the data table for the axial stiffness of Takabatake et al. [24].

When using the formula of Mizutani et al. [19], the impulse duration of ASCE [13] was used for dt.

The calculation results are shown in Figure 12. The formula of the Bridge and Engineering Association [14] gives too big values as compared with the mean values. The formula of the Japan Road Association [16] gives small values as compared with the mean values of all formulae.

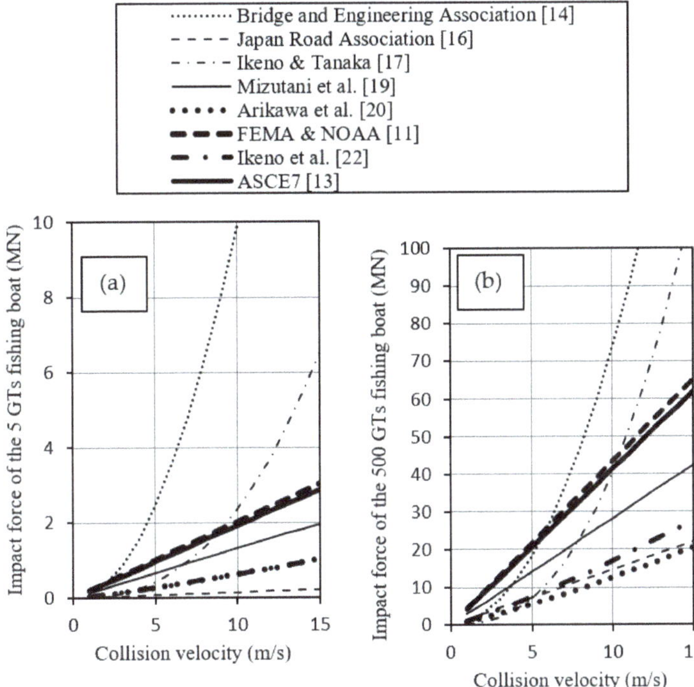

Figure 12. Impact force of the fishing boat: (**a**) the case of the 5 GTs boat, and (**b**) the case of the 500 GTs boat.

3. Reliability Examination of Main Formulae

Concerning the calculated and measured values of the impact force in Section 2.2, we classified formulae, which considerably overestimate or underestimate as "×", formulae, which overestimate moderately or give values near the measured values as "○" and formulae, which give intermediate values of these formulae as "△". Moreover, in examining the range of the calculated values in Section 2.3, since true values are not known, we classified formulae, which give too large or too small values compared with the mean values of all formulae as "×". Then, we judged that only formulae classified as "○" in Section 2.2 and were not classified as "×" in Section 2.3 should be used because their reliability is high, and formulae, which were classified as "×" in both sections should not be used. These results are shown in Table 4.

Table 4. Results of the reliability examination of impact force formulae.

Driftage	Examination Type	Bridge and Eng. Assoc. [14]	Matsutomi [15]	Japan Road Assoc. [16]	Ikeno and Tanaka [17]	Mizutani et al. [18,19]	Arikawa et al. [20]	FEMA and NOAA [11]	Ikeno et al. [22]	ASCE [13]
Driftwood	Correlation		○	×	×		△	○	△	○
	Existence range				×					
	Reliability		○		×			○		○
Container	Correlation		○	×	×	△	○	○	○	○
	Existence range			×						
	Reliability		○	×			○	○	○	○
Car	Correlation	△		×	×	△	×	○	×	○
	Existence range			×						
	Reliability			×				○		○
Fishing Boat	Correlation	△		△	×	○	△	○	△	○
	Existence range	×		×						
	Reliability					○		○		○

(○ indicates "good", △ indicates "ordinary", and × indicates "bad").

The following can be concluded based on Table 4:

(1)   In the case of driftwood, the calculation formulae of Matsutomi [15], FEMA and NOAA [11] and ASCE [13] are most reliable.

(2)   In the case of containers, the calculation formulae of Matsutomi [15], Arikawa et al. [20], FEMA and NOAA [11], Ikeno et al. [22], and ASCE [13] are most reliable.

(3)   In the case of cars, the calculation formulae of FEMA and NOAA [11] and ASCE [13] are the most reliable.

(4)   In the case of ships, the calculation formulae of Mizutani [19], FEMA and NOAA [11], and ASCE [13] are the most reliable.

However, when FEMA and NOAA [11] and ASCE [13] are used, although the appropriate stiffness is very important, it is very difficult to estimate the appropriate stiffness of complicated structures. Moreover, the stiffness changes according to the change of collision velocity. Therefore, we used the stiffness recommended by FEMA and NOAA [11] and ASCE [13] in the case of a 6.1 m container and a 12.2 m container, and the recommended stiffness of Takabatake et al. [24] or Kaida and Kihara [6] for cars and ships.

Moreover, since the formulae of the Bridge and Engineering Association [14], Matsutomi [15], Mizutani et al. [18,19], Arikawa et al. [20], and Ikeno et al. [22] show good correlations with the measured values, if the formulae were improved so that parameters concerning the rigidity of the driftage and the collided body could be considered, the reliability of the formulae becomes better, and the applicability range also increases.

On the other hand, since the formula of the Japan Road Association [16] cannot make the difference in rigidity of the driftage and the collided body into consideration and tends to give an excessive impact force, it is better to avoid its use. In addition, since the formula of Ikeno and Tanaka [17] can also not make the difference in rigidity of the driftage and the collided body into consideration and tends to give a too small impact force, it is better to avoid its use.

4. Conclusions

There are many formulae for calculating the impact forces from driftage. Since there are considerable differences between the calculated values of these formulae, the reliability of each formula was assessed here. We summarized the main calculation formulae of the impact force proposed by 2019 and compared the calculation values of these formulae with the measured data from large-scale experiments. Furthermore, we checked the range of the calculation values of each formula up to 15 m/s in collision velocity and then clarified the following points:

(1)   In the case of driftwood, the formulae of Matsutomi [15], FEMA and NOAA [11], and ASCE [13] are the most reliable;

(2)   In the case of containers, the formulae of Matsutomi [15], Arikawa et al. [20], FEMA and NOAA [11], Ikeno et al. [22], and ASCE [13] are most reliable;

(3)   In the case of cars, the formulae of FEMA and NOAA [11] and ASCE [13] are the most reliable;

(4)   In the case of ships, the formulae of Mizutani [19], FEMA and NOAA [11], and ASCE [13] are the most reliable.

Here, except for the formulae of the Japan Road Association [16] and Ikeno and Tanaka [17], the stiffness or parameters governed by rigidity (stop distance of ships, yield stress, collision time, elasticity modulus) must be set appropriately:

(a)   When FEMA and NOAA [11] and ASCE [13] are used, the stiffness recommended by FEMA and NOAA [11] and ASCE [13] can be used for the 6.1 m container and the 12.2 m container, and the stiffness recommended by Takabatake et al. [24] or Kaida and Kihara [6] can be used for cars and ships;

(b)   Since the formulae of the Bridge and Engineering Association [14], Matsutomi [15], Mizutani et al. [18,19], Arikawa et al. [20], and Ikeno et al. [22] show good correlations with the measured values, provided the formulae are improved so that parameters concerning the rigidity of the driftage and the collided body can be considered, the reliability of the formulae is expected to increase, and the applicable scope to become larger;

(c)   Since the formulae of the Japan Road Association [16] and Ikeno and Tanaka [17] cannot make the difference in rigidity of the driftage and the collided body into consideration, the applicable range of their formulae is limited. Moreover, the former tends to give an excessive impact force, and the latter tends to give a too small impact force. Therefore, it is desirable to avoid using their equations.

Author Contributions: Conceptualization, Y.Y. and M.T.; methodology, Y.Y. and M.T.; software (Excel sheets for calculating the impact forces), Y.Y., Y.K., Y.N. and M.T.; validation, Y.Y. and M.T.; formal analysis, all members; investigation, all members; resources, all members; data curation, Y.Y.; writing—original draft preparation, Y.Y.; writing—review and editing, Y.Y, E.M.; visualization, supervision, Y.Y.; project administration, M.T. All authors have read and agreed to the published version of the manuscript.

Funding: This research was partly funded by the grant-in-aid for scientific research (c; 18k04667) of JSPS and the Core Research Cluster of Disaster Science at Tohoku University. We express our deepest gratitude.

Institutional Review Board Statement: This research dose not involve humans or animals.

Informed Consent Statement: This research dose not involve humans.

Data Availability Statement: In this research, all photos were taken by authors and all data used for calculations are open data from the references.

Conflicts of Interest: The authors declare no conflict of interest.

References

1.   Nistor, I.; Goseberg, N.; Stolle, J. Tsunami-Driven Debris Motion and Loads: A Critical Review. *Front. Built Environ.* **2017**, *3*. [CrossRef]

2.   Naito, C.; Cercone, C.; Riggs, H.R.; Cox, D. Procedure for Site Assessment of the Potential for Tsunami Debris Impact. *J. Waterw. Port Coastal Ocean Eng.* **2014**, *140*, 223–232. [CrossRef]

3.   Yoneyama, N.; Tanaka, Y.; Pringle, J.W.; Nagashima, H. The development of three dimensional numerical analysis for tsu-nami driven debris in real scale senarios and its basic verification. *J. JSCE B2 (Coast. Eng.)* **2015**, *71*, 1027–1032. (In Japanese)

4.   Murase, F.; Teramoto, N.; Toyoda, A.; Tanaka, Y.; Arikawa, T. Examination for Evaluation Method of Collision Probability of Tsunami Debris Using Numerical Calculation. *J. Jpn. Soc. Civ. Eng. Ser. B2 (Coast. Eng.)* **2019**, *75*, I_439–I_444. [CrossRef]

5.   Magoshi, K.; Ge, H.; Nonaka, T.; Harada, T.; Murakami, K. Collision Simulation of A Large Flotsam and A Long-Span Bridge in A Tsunami. *J. Jpn. Soc. Civ. Eng. Ser. B3 (Ocean Eng.)* **2012**, *68*, 222–227. [CrossRef]

6.   Kaida, H.; Kihara, N. *Evaluation Technologies for the Impact Assessment of Tsunami Debris on Nuclear Power Plants-Review of Current Status and Discussion on the Application*; Central Research Institute of Electric Power Industry: Tokyo, Japan, 2017; Volume o16010, 45p. (In Japanese)

7.   Stolle, J.; Derschum, C.; Goseberg, N.; Nistor, I.; Petriu, E. Debris impact under extreme hydrodynamic conditions part 2: Impact force responses for non-rigid debris collisions. *Coast. Eng.* **2018**, *141*, 107–118. [CrossRef]

8.   Stolle, J.; Nistor, I.; Goseberg, N.; Petriu, E. Multiple Debris Impact Loads in Extreme Hydrodynamic Conditions. *J. Waterw. Port Coast. Ocean Eng.* **2020**, *146*, 04019038. [CrossRef]

9.   Haehnel, R.; Daly, S.F. Maximum impact force of woody debris on floodplain structures. *J. Hydraul. Eng.* **2004**, *130*, 112–120. [CrossRef]

10.   ASCE. *ASCE/SEI, Standard 7-10, Minimum Design Loads for Buildings and Other Structures*; ASCE: Virginia, WA, USA, 2010.

11.   FEMA. *Guidelines for Design of Structures for Vertical Evacuation from Tsunamis*, 2nd ed.; FEMA: Washington, DC, USA, 2012.

12.   Aghl, P.P.; Naito, C.J.; Riggs, H.R. Full-Scale Experimental Study of Impact Demands Resulting from High Mass, Low Velocity Debris. *J. Struct. Eng.* **2014**, *140*, 04014006. [CrossRef]

13.   ASCE. *Standard 7, Chapter 6, Tsunami Loads and Effects*; ASCE: Virginia, WA, USA, 2015.

14.   Marine Bridge Research Committee. *Honshu Shikoku Bridge Technical Standards, Lower Volume, Sub-Structure Design Standards and Explanations*; Bridge and Engineering Association: Tokyo, Japan, 1978. (In Japanese)

15.   Matsutomi, H. A practical formula for estimating impulsive force due to driftwoods and variation features of the impulsive force. *Doboku Gakkai Ronbunshu* **1999**, *621*, 111–127. (In Japanese) [CrossRef]

16.   Japan Road Association. *Specifications for a Highway Bridge and Explanations, Part I Common*; Japan Road Association: Tokyo, Japan, 2002; pp. 65–67. (In Japanese)

17. Ikeno, M.; Tanaka, H. Experimental study on impulse force of drift body and tsunami running up to land. *JSCE. Coast. Eng.* **2003**, *50*, 721–725. (In Japanese)
18. Mizutani, N.; Takagi, Y.; Shiraishi, K.; Miyajima, S.; Tomita, T. Study on wave force on a container on apron due to tsunamis and collision force of drifted container. *Annu. J. Coast. Eng. JSCE* **2005**, *52*, 741–745. (In Japanese)
19. Mizutani, N.; Usami, A.; Koike, T. Experimental study on behavior of drifting boats due to tsunami and their collision forces. *Annu. J. Civ. Eng. ocean, JSCE* **2007**, *23*, 63–68. (In Japanese)
20. Arikawa, T.; Ohtsubo, D.; Nakano, F.; Shimosako, K.; Ishikawa, N. Large Model Tests of Drifting Container Impact Force due to Surge Front Tsunami. *Proc. Coast. Eng. JSCE* **2007**, *54*, 846–850. [CrossRef]
21. Arikawa, T.; Washizaki, M. Large Scale Tests on Concrete Wall Destruction by Tsunami with Driftwood. *J. Jpn. Soc. Civ. Eng. Ser. B2 Coast. Eng.* **2010**, *66*, 781–785. [CrossRef]
22. Ikeno, M.; Kihara, N.; Takabatake, D. Simple and practical estimation of movement possibility and collision force of de-bris due to tsunami. *J. JSCE B2 Coast. Eng.* **2013**, *69*, 861–865. (In Japanese)
23. Ikeno, M.; Takabatake, D.; Kihara, N.; Kaida, H.; Miyagawa, Y.; Shibayama, A. Collision Experiment of Woody Debris and Improvement of Collision Force Formula. *J. Jpn. Soc. Civ. Eng. Ser. B2 Coast. Eng.* **2015**, *71*, 1021–1026. [CrossRef]
24. Takabatake, D.; Kihara, N.; Miyagawa, Y.; Kaida, H.; Shibayama, A.; Ikeno, M. Axial Stiffness Model for Estimating Floating Debris Impact Forces due to Tsunami. *J. Jpn. Soc. Civ. Eng. Ser. B2 Coast. Eng.* **2015**, *71*, 1015–1020. [CrossRef]
25. Arita, K. A study on the strength of ships and other structures against collision. *Rep. Natl. Marit. Res. Inst.* **1988**, *25*, 35–125. (In Japanese)

MDPI

St. Alban-Anlage 66

4052 Basel

Switzerland

Tel. +41 61 683 77 34

Fax +41 61 302 89 18

www.mdpi.com

Journal of Marine Science and Engineering Editorial Office

E-mail: jmse@mdpi.com

www.mdpi.com/journal/jmse